A Novel Method for Predicting Product Properties
in Fluidized Bed Spray Granulation

A Novel Method for Predicting Product Properties in Fluidized Bed Spray Granulation

Vom Promotionsausschuss der
Technischen Universität Hamburg

zur Erlangung des akademischen Grades

Doktor-Ingenieur (Dr.-Ing.)

genehmigte Dissertation

von

Paul Kieckhefen

aus
Hamburg

2021

Bibliografische Information der Deutschen Nationalbibliothek
Die Deutsche Nationalbibliothek verzeichnet diese Publikation in der Deutschen Nationalbibliografie; detaillierte bibliographische Daten sind im Internet über http://dnb.d-nb.de abrufbar.
1. Aufl. - Göttingen: Cuvillier, 2021
Zugl.: (TU) Hamburg, Univ., Diss., 2021

A Novel Method for Predicting Product Properties in Fluidized Bed Spray Granulation
Date of Examination: 16.12.2021
Reviewers:
Prof. Dr.-Ing. habil. Dr. h.c. Stefan Heinrich
Prof. Dr.ir. J.A.M. Kuipers
Supervisor: Prof. Dr.-Ing. habil. Dr. h.c. Stefan Heinrich

Institute of Solids Process Engineering and Particle Technology
Hamburg University of Technology
Denickestraße 15
21075 Hamburg
Germany

Nonnenstieg 8, 37075 Göttingen
Telefon: 0551-54724-0
Telefax: 0551-54724-21
www.cuvillier.de

1. Auflage, 2021
Gedruckt auf umweltfreundlichem, säurefreiem Papier aus nachhaltiger Forstwirtschaft.

ISBN 978-3-7369-7556-9
eISBN 978-3-7369-6556-0

Abstract

The objective of this work is to further develop the usability of simulations for the product-property oriented design of fluidized bed spray granulators. In this process, a solids-containing liquid is introduced into a fluidized bed of core particles which grow in size either by layering by the solids contained in the liquid or agglomerating due to adhesive forces. Scaling this class of process poses a great challenge due to the dependence of the formed product's properties on the wetting and drying conditions. These differ substantially between laboratory trials, in which these processes are developed, and the production scale due to dissimilar mixing patterns. The coupled Computational Fluid Dynamics-Discrete Element Method (CFD-DEM), the state-of-the-art in the simulation of fluid and particle dynamics in fluidized bed systems, can accurately describe wetting and drying, as well as mixing patterns. Thus, this method was used for further developments.

To lay the foundation for a successful prediction whether agglomeration or layering occurs, an approach to treat weakly wetted granular systems in the Discrete Element Method was developed. This one-parameter liquid bridge state model was demonstrated to reproduce a wide variety of granular flow situations with low liquid loading. An efficient segregated workflow for the calibration of this and the liquid bridge force models is presented.

To predict the surface roughness in layering granulation, the concept of tracked quantities was developed. These are recorded during the course of a simulation for structure-determining events, such as droplet impacts or the drying of liquid on a particle's surface. By paralleling granulation experiments and these simulations, a mapping between tracked quantities and product properties can be obtained. This mapping can then be applied to simulation cases at different scales and with differing mixing patterns to predict the product property in question. The approach was demonstrated to be able to accurately describe both liquids containing a dissolved salt and a suspension. Finally, the ability to quantitatively predict the change in product properties during scale-up to pilot-scale was shown even under variation of the number of nozzles.

Zusammenfassung

Ziel dieser Arbeit ist die Erschließung von Simulationsmethoden zur gezielten Auslegung der Wirbelschichtsprühgranulation unter Beachtung von Produkteigenschaften. In diesem Prozess wird eine Feststoff-enthaltende Flüssigkeit in ein fluidisiertes Partikelbett eingesprüht, wobei die Kernpartikel entweder durch Ablagerung des Feststoffs in der Flüssigkeit in Schichten wachsen oder durch Haftkräfte agglomerieren. Die Maßstabsübertragung dieser Prozesse stellt eine große Herausforderung durch die Abhängigkeit der Eigenschaften des gebildeteten Produktes auf die Befeuchtungs- und Trocknungsbedingungen im Apparat dar. Diese unterscheiden sich deutlich zwischen Labormaßstab, in dem die Produktformulierungen entwickelt werden, und dem Produktionsmaßstab in Folge von abweichenden Mischungsmustern. Die gekoppelte Numerische Strömungsmechanik-Diskrete-Elemente-Methode (Computational Fluid Dynamics-Discrete Element Method, CFD-DEM), die Stand der Technik in der Wirbelschichtsimulation ist, kann Befeuchtung und Trocknung ebenso wie Durchmischung deterministisch beschreiben und wurde für die Modellentwicklung in dieser Arbeit eingesetzt.

Um eine Grundlage für die Unterscheidung zwischen den beiden genannten Wachstumsmechanismen zu schaffen, konnte ein Ansatz zur Simulation gering benetzter Schüttungen mit der Diskrete-Elemente-Methode entwickelt werden. Die Fähigkeit des Ein-Parameter Flüssigkeitsbrücken-Zustandsmodells, viele Schüttungszustände realitätsgetreu zu beschreiben, wurde für ein Spektrum an Befeuchtungszuständen gezeigt. Ein effizienter, entkoppelter Arbeitsablauf zur Kalibrierung dieses Modells und der Flüssigkeitsbrücken-Kraftmodelle wird vorgestellt und an einem Beispielsystem demonstriert.

Zur Vorhersage der Oberflächenrauheit in der Schichtaufbaugranulation wurde das *tracked quantities*-Konzept entwickelt. Diese werden während der Simulation für strukturformende Ereignisse, wie Tropfenabscheidung oder Abtrocknung der Oberflächenflüssigkeit, aufgezeichnet. Durch die Durchführung von Zwillingssimulationen für einen Satz an Experimenten kann eine Abbildung von den *tracked quantities* auf die Eigenschaften des Produktes erhalten werden. Diese Abbildung kann dann auf Simulationen von Anlagenentwürfen auf den verschiensten Skalen angewendet werden, um den Einfluss der beschriebenen Faktoren auf die Produkteigenschaften zu erfassen. Der Ansatz konnte die Testfälle der Granulation von Suspensionen und Salzlösungen erfolgreich abbilden. Schließlich konnte gezeigt werden, dass mit der Methode in eine Maßstabsübertragung vom Labor- auf den Pilotmaßstab unter Variation der Düsen im System die Produkteigenschaften quantitativ vorhergesagt werden können.

Acknowledgements

Throughout the writing of this dissertation I have received a great deal of support and assistance. I would like to extend my sincerest thanks and gratitude to the following individuals.

I would first like to thank my supervisor, Prof. Stefan Heinrich, whose expertise in research and leadership made my time at SPE a fruitful one. His skill and experience with leading projects helped me refine my work throughout the past few years and develop into an independent researcher. He encouraged heavy collaboration at SPE, allowing me to cooperate with a wide range of researchers and companies.

I would also like to thank Prof. Hans Kuipers for critiquing my thesis in great detail and for carefully examining my work, both before and during my defense. Furthermore, Prof. Stefan Pirker and Dr. Thomas Lichtenegger have served as mentors, offering invaluable advice, inspiring me to pursue my research interests and demonstrating how to keep my work to a high standard.

Next, I would like to thank BASF for funding the project. Without their contribution, none of this work would have been possible. I would especially like to thank my BASF liasons Moritz Höfert and Dominik Weis for their mentorship and for always keeping my work on track. Their consistent input and valuable feedback kept my work focused.

I would also like to thank my colleagues at SPE for their continuous support in work and play, in the office and at conferences. Their collaboration allowed me to expand my research interests to many different areas of specialization. I owe particular thanks to Dr. Swantje Pietsch-Braune who took me on as a Bachelor student and has been there all these years as an advisor that always reviewed my work with an attentive eye. She has witnessed and heavily supported my progression from an earnest Bachelor student to an autonomous researcher. I also want to thank my office mates Anna Schütt and Maike Orth for putting up with me and commiserating about the life as a doctoral researcher.

I am particularly indebted to my student research assistants, Christoph Wolter and Brig Watson, for supporting the experimental part of my work with their hard work, persistence and enthusiasm. I also owe great thanks to Philipp Grohn, who has contributed his expertise and skill in nozzle characterization.

Furthermore, I would like to support my girlfriend Deifilia To for proof-reading this work and supporting me through the ups and downs of research.

Lastly, I would like to thank my parents Nina and Ulrich and my sister, Ida. My family has supported and cheered me on throughout my studies and research over the last ten years. The completion of this dissertation would not have been possible without them.

Contents

List of Symbols

Greek Symbols

Symbol	Description	Unit
α	Phase fraction	$m^3_{phase}\ m^{-3}_{total}$
α	Heat transfer coefficient	$m^3_{phase}\ m^{-3}_{total}$
β	Scaling parameter	–
β	Momentum exchange coefficient	$N\,s\,m^{-1}$
β	Mass transfer coefficient	$kg\,s\,m^{-1}\,m^{-3}$
γ	Damping coefficient	$N\,m^{-1}$
γ_{rab}	Model coefficient	$N\,m^{-1}$
δ	Overlap	m
δ_{lb}	Liquid bridge length	m
δ_{CG}	Coarse-graining scaling factor	–
Δ	Difference / step	–
η	Drying potential	–
η	Efficiency	–
η_i	Dynamic viscosity of phase *i*	Pa s
θ	Angle	rad
λ	Filter coefficient	m^{-1}
μ	Friction coefficient	–
σ	Normal stress	Pa
τ	Shear stress	Pa
τ	Time constant	s
τ	Viscous stress tensor	Pa
ψ	Fractional surface coverage	–
ϕ	Mass flux	$kg\,s^{-1}\,m^{-2}$
ω	Rotational velocity	$rad\,s^{-1}$

Roman Symbols

Symbol	Description	Unit
A	Area	m^2
A	Model coefficient	–
B	Model coefficient	–
C	Model coefficient	–
d	Diameter	m
e	Specific energy	$J\,m^{-3}$
$\mathbf{e}$	Unit vector	–
E	Energy	J
f	Area fraction	$m^2\,m^{-2}$
$\mathbf{F}$	Force	N
F	Nondimensionalized drag force	–
$\mathbf{g}$	Gravitational acceleration	$m\,s^{-2}$
G	Shear modulus	Pa
h	Height	m
J	Moment of inertia	$kg\,m^2$
J	Objective function	–
k	Turbulent kinetic energy density	$J\,m^{-3}$
k	Spring stiffness	$N\,m^{-1}$
K	Kinetic energy density	$J\,m^{-3}$
M	Mass	kg
N	Number	–
p	Pressure	Pa
Q	Heat transfer rate	$J\,s^{-1}\,m^{-3}$
r	Radius	m
S	Stiffness	$N\,m^{-1}$
S	Surface area	m^2
S	Source term	–
t	Time	s
$\mathbf{T}$	Torque	N m
T	Temperature	K or °C
V	Volume	m^3
x	Length	m
$\mathbf{x}$	Position of a particle	m
X	Solid loading	$kg\,kg_{bulk}^{-1}$
y_i	Mass fraction of species i in the fluid phase	$kg\,kg^{-1}$

Symbol	Description	Unit
Y	Young's modulus	Pa
Y	Water vapor mass loading	$kg_{vapor}\ kg_{dry\ gas}^{-1}$

Subscripts/Superscripts

Symbol	Description
*	Reduced
*	At interface
cap	Regarding the capillary force
cell	Regarding a single cell of the CFD mesh
comb	Combustion
d	Droplet
drag	Drag force
dep	Deposition
e	Energy
evap	Evaporation
ex	Explicit
exp	Experiment
ext	External
f	Regarding the fluid phase
g	Regarding the gas phase
im	Implicit
in	Entering the system
k	Elastic
l	Liquid
lb	Liquid bridge
min	Minimum
n	In the normal direction
out	Leaving the system
p	Regarding the particle phase
proj	Projected
r	Relative
rab	Regarding the Rabinovich law
red	Reduced
[illegible]	Static friction
rfr	Rolling friction

Symbol	Description
s	Regarding the solid phase
sim	Simulation
soulie	Regarding the Soulié law
spray	Regarding the spray
surface	Regarding the surface
t	In the tangential direction
u	Regarding the momentum equation
v	Constant volume
vap	Vapor phase
w	Water (liquid)
w	Wall
wet	Wetted
wb	Wet bulb state

Abbreviations

Abbreviation	Description
CFD	Computational Fluid Dynamics
DEM	Discrete Element Method

1 Introduction

Fluidized bed spray granulation is a process in which particle growth is induced through the injection of solids in liquid form into a granulator, either as a melt, or suspended or dissolved in a liquid. The process is illustrated in Fig. 1.1. Here, a particle bed is fluidized using the inflowing air and thereby mixed. Droplets are injected through a nozzle and deposited on the particles. After deposition, the liquid interacts with the gas phase. In the case of melt granulation, the gas acts as a cooling agent and forces the solids to solidify whereas in the case of solution or suspension granulation, the liquid evaporates. In this work, the focus is laid on the treatment of solutions or suspension rather than melt granulation.

Fluidized bed spray granulation is commonly chosen for its consistency in producing products with sharply defined properties or functionality. It is thus widely used in the food industry, pharmaceutical industry as well as consumer product and chemical industries. These resulting product properties are the consequence of the local conditions that particles experience over the course of the process. Local process conditions include the deposition of spray droplets onto the particle surface, contact with other particles and evaporation of liquid on the surface, as well as heat transfer with the gas phase. The fluidization gas introduces agitation into the system and allows for evaporation of the liquid. The physics

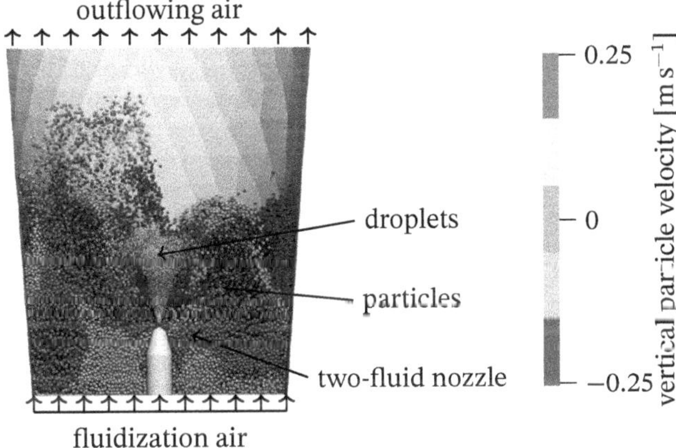

Fig. 1.1.: Snapshot of a CFD-DEM simulation of the particle and droplet dynamics inside a Glatt GF3 granulator in operation, with particles clipped away up to the mid-plane.

of fluidization allow for excellent heat and mass transfer due the intense interaction of these two phases.

Depending on the properties and amount of spray liquid applied and the velocities at which particles collide, they may either grow in layers due to the solids contained in the liquid, or by particles agglomerating together. Layer growth occurs for high velocity impacts, spray liquids with a low surface tension and viscosity, and low spray rates (equating to low local liquid concentrations). Here, contacts of wetted particles will not lead to sticking. This is both because wetted contacts rare occur, and the contacts endure for durations that are too short to result in solidification or sintering (Salman et al., 2006).

Hoffmann et al. (2015) and Diez et al. (2018) identified the global drying potential

$$\eta = \frac{Y_{\mathrm{in}}^{\mathrm{wb}} - Y_{\mathrm{out}}}{Y_{\mathrm{in}}^{\mathrm{wb}} - Y_{\mathrm{in}}} \tag{1.1}$$

to correlate with a wide range of layering granulation product properties such as porosity, yield strength and surface roughness when spraying a solution of a salt. Y is the specific humidity loading and $Y_{\mathrm{in}}^{\mathrm{wb}}$ the wet-bulb humidity that would occur in the granulator if the enthalpy of the inflowing air is used for evaporation until the saturation point is reached. Values approaching $\eta = 1$ indicate low spray rates or high inflowing gas temperatures because the energy budget of the inflowing air is large compared to rate at which liquid is evaporated. In contrast, values close to $\eta = 0$ indicate low gas temperatures and high spray rates, indicating that the outflowing air already reached its saturation point and no further evaporation can occur. The drying potential is an indirect descriptor for the rate at which the droplets and wetted particles dry in the granulator, showing that these are key factors in granulation. However, correlating the drying potential with product properties fails when spraying suspensions (Schmidt et al., 2017b). Another important branch of layering spray granulation is the use of melts, for which no quantitative morphology studies could be found in literature at the time of this work.

The other process, agglomeration, occurs when the spray liquid and process conditions are tailored to let particle collisions lead to sticking and solidification. Particular focus is laid on the strength of liquid bridges between two colliding particles. The stability of the process is dependent on an equilibrium between solid bridges forming and breaking. When bridges, that are much stronger than the stresses in the system, form, unbounded agglomerate growth occurs, causing fluidization to cease. On the other hand, bridges that are weaker than the stresses in the system would break, leading to the prior case of layer growth.

Fluidization, the key phenomenon at the heart of fluidized bed granulation, can nowadays be accurately described with numerical methods, most notably and suitably the coupled Computational Fluid Dynamics-Discrete Element Method (CFD-DEM), as described by Tsuji et al. (1992). The trajectories of up to tens of millions of individual particles (or parcels) can be resolved with limited effort and their exchange of heat and mass with the gas phase can be tracked. Despite that, fully space-resolved simulations where the structure of particles themselves is resolved still pose a substantial challenge. Therefore, a variety of indirect approaches have been developed to use simulation technologies.

Terrazas-Velarde et al. (2009) used a Monte Carlo approach to describe agglomeration in fluidized bed spray granulation by tracking discrete, stochastic events. Based on this, Dadkhah et al. (2012) were able to accurately model agglomerate structure. The weakness of Monte Carlo approaches lie in the need to provide probabilities for events and other closures, such as for the collision velocity of particles. Some works, such as those of Dosta et al. (2013), have developed elaborate ways to bridge length- and time scales by deriving closures from CFD-DEM simulations. Kafui and Thornton (2008) directly described agglomerate growth using CFD-DEM simulations with cohesion / surface-energy modelling, albeit for a very small system. Fries et al. (2014) used the CFD-DEM method to investigate the resulting properties of fluidized bed spray agglomeration. They were successful in predicting agglomerate breakage strength by using the aforementioned growth-breakage equilibrium for a variety of different granulator geometries without having to rely on a wide variety of closures.

Few studies have been performed with respect to product properties of fluidized bed layering spray granulation. Hoffmann (2016) used experimental closures based on the global drying potential to describe the growth of particles using a population balance model. For the surface structures formed by layering granulation, Jiang et al. (2020) performed CFD-DEM simulations coupled to a Monte Carlo method with discretized particle surfaces to characterize the inter-particle homogeneity of layer formation, albeit with no correlation to the actual particle porosity.

1.1. Aim of this Work

The aim of this work is the maturation of simulation technology for performing predictive simulations of the fluidized bed spray granulation process at any given apparatus scale. To this end, two focus points were chosen:

- Simulation of weakly wetted granular matter.
- Prediction of fluidized bed spray layering granulation product properties.

Both aims are independent of each other for the purpose of this thesis. The ability to model the influence of liquid distribution on dynamics allows to determine whether agglomeration or layering granulation will take place. Finding a calibration approach to describe weakly wetted, sensitive systems is therefore one of the aims of this thesis.

The other focus is the development and application of an approach for the prediction of fluidized bed spray layering granulation using CFD-DEM simulations, leveraging knowledge of the conditions under which drying occurs. To this end, an approach is to be developed that

- works for spray liquids that are solutions or suspensions and
- can predict deviations in product properties that occur in scale-up.

This is to be done in a manner that is easy and reproducible.

1.2. Structure of this Work

Chapter 2 gives an overview of the classical Discrete Element Method and Computational Fluid Dynamics techniques and their coupling. This provides the theoretical foundation that is later built upon.

The succeeding content chapters are split into two parts. The first part addresses the aim of describing weakly wetted granular systems. In chapter 3, an overview of the physics involved in liquid bridges is given, a model is composed and a new liquid bridge state model is proposed. The sensitivities of characterization experiments with respect to the liquid bridge model parameters are analyzed and a calibration workflow is proposed on the basis of these specific sensitivities. The approach is applied to a real material system in chapter 4.

For the second part, a product-property prediction approach for fluidized bed layering spray granulation is devised in chapter 6. The applicability of this approach to the use of suspensions as the spray liquid is shown in chapter 7 and the applicability to using salt solutions as the spray liquid is shown in chapter 8. This is done by demonstrating the ability to find a mapping between the conditions that particles experience and the product property chosen. Chapter 9 then applies the mapping derived in chapter 8 to a pilot-scale case and demonstrates the predictiveness of the approach in scale-up by a factor of 8.

2 Numerical Simulation of Particulate Flows using CFD-DEM

The CFD-DEM (Computational Fluid Dynamics-Discrete Element Method) is a versatile, deterministic method of simulating fluid-solid flows due to its ability to represent both the gaseous flow and the solid particles in their most natural frames of reference respectively. An expansive overview of the range of applications it is used for can be found in Kieckhefen et al. (2020), from which this chapter was derived.

In this chapter, a short description of the individual DEM and CFD methods are given, followed by an overview of the details that must be considered when coupling these methods.

2.1. Governing Equations of the Discrete Element Method

In the Discrete Element Method (DEM), as proposed by Cundall and Strack (1979), particle motion is resolved by numerically integrating the Newtonian equations of motion:

$$\ddot{\mathbf{x}}_i = \frac{1}{M_i}\left(\sum_j \mathbf{F}_{j\to i} + \mathbf{F}_{i,\mathrm{ext}}\right) \tag{2.1}$$

$$\dot{\omega}_i = \frac{1}{J_i}\left(\sum_j \mathbf{T}_{j\to i} + T_{i,\mathrm{ext}}\right) \tag{2.2}$$

where $\mathbf{x}_i$ is the position of a particle i, M_i refers to its mass and $\mathbf{F}_{i,\mathrm{ext}}$ to the external forces acting upon it, such as inter-phase drag and gravity. ω_i is the angular momentum, J_i the moment of inertia and $\mathbf{T}_{j\to i}$ the torque acting on i due to j. The inter-particle contact force $\mathbf{F}_{j\to i} = \mathbf{F}^{\mathrm{n}}_{j\to i} + \mathbf{F}^{\mathrm{t}}_{j\to i}$ of particle j acting upon particle i are composed of a normal component $\mathbf{F}^{\mathrm{n}}_{j\to i}$ and a tangential component $\mathbf{F}^{\mathrm{t}}_{j\to i}$.

There are two different approaches to solving these equations of motion: the soft-sphere and the hard-sphere model. The soft-sphere model uses a global time step and performs time-integration over the sum of all forces acting upon the particles. Particles are allowed to overlap and contact forces are modeled, for example, using the Hertz-Mindlin model, as illustrated in Fig. 2.1. These overlap-dependent force models take into account both single-particle mechanical properties such as the coefficient of restitution, the elasticity modulus and Poisson ratio, as well as frictional bulk properties. The maximum particle overlap should be kept under a value of 0.3% of the particle radius (Lommen et al., 2014) for numerical stability. This stability criterion restricts the global time step.

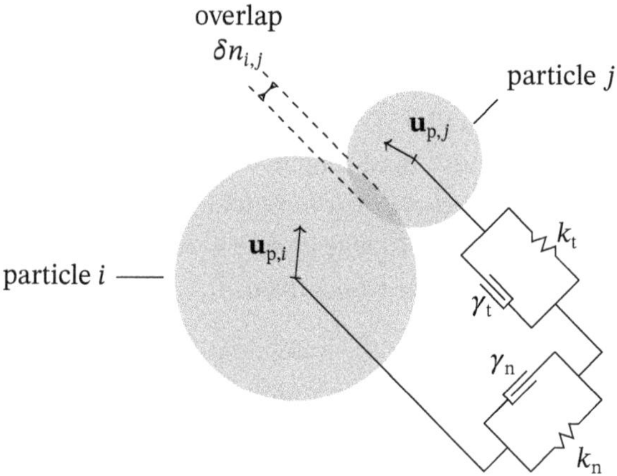

Fig. 2.1.: Schematic of particle overlap in the soft-sphere Discrete Element Method (DEM). Springs are indicated by the symbol for their coefficient k and dashpots are indicated by their viscous coefficient γ.

The hard-sphere model follows a localized approach to time stepping, where particle tracks are assumed to be linear or linearly accelerated/dampened until a collision is detected. Then, the coefficient of restitution is applied to the colliding elements, the deflection is applied to the velocity vectors and calculation resumes. The operation is periodically interrupted to recalculate non-contact forces.

While the effective timestep is much lower in soft-sphere models, hard-sphere models are rarely chosen for bulk solids due to the high incidence of particle contacts and a high collision frequency. As every collision requires an interruption of calculation, the hard-sphere model is rarely used for these cases.

In soft-sphere simulations, contact forces are calculated according to the Hertz law in the normal direction (Kloss et al., 2012):

$$\mathbf{F}^{\mathrm{n}}_{j\to i} = k_{\mathrm{n}}\,\delta\mathbf{n}_{ij} - \gamma_{\mathrm{n}}\mathbf{u}^{\mathrm{n}}_{\mathrm{p},ij} \tag{2.3}$$

$$\mathbf{F}^{\mathrm{t}}_{j\to i} = k_{\mathrm{t}}\,\delta\mathbf{t}_{ij} - \gamma_{\mathrm{t}}\mathbf{u}^{\mathrm{t}}_{\mathrm{p},ij} \tag{2.4}$$

where k is the elastic coefficient, $\delta\mathbf{n}_{ij}$ and $\delta\mathbf{t}_{ij}$ are the normal and tangential overlaps, γ is the damping coefficient and $\mathbf{u}_{\mathrm{p},ij}$ is the relative velocity.

These coefficients can be calculated from bulk solids parameters such as the particle mass m_i, the radius R_i, the modulus of elasticity Y_i and the coefficient of restitution e using the model equations given in Tab. 2.1.

Tab. 2.1.: Hertz-Mindlin-Tsuji contact model quantities.

Quantitiy	Expression
Spring stiffness	
Normal	$k_{\mathrm{n}} = (4/3)Y^*\sqrt{r^*\delta n_{ij}}$
Tangential	$k_{\mathrm{t}} = 8G^*\sqrt{r^*\delta n_{ij}}$
Damping coefficient	
Normal	$\gamma_{\mathrm{n}} = -2\sqrt{5/6}\beta\sqrt{S_{\mathrm{n}}m^*} \geq 0$
Tangential	$\gamma_{\mathrm{t}} = -2\sqrt{5/6}\beta\sqrt{S_{\mathrm{t}}m^*} \geq 0$
Stiffness	
Normal	$S_{\mathrm{n}} = 2Y^*\sqrt{r^*\delta n_{ij}}$
Tangential	$S_{\mathrm{t}} = 8G^*\sqrt{r^*\delta n_{ij}}$
Scaling parameter	$\beta = \ln(e)/\sqrt{\ln^2(e) + \pi^2}$
Effective quantities	
Young's modulus	$Y^* = 1/\left(\left(1-\nu_1^2\right)/Y_1 + \left(1-\nu_2^2\right)/Y_2\right)$
Shear modulus	$G^* = 1/\left(2(2-\nu_1)(1+\nu_1)/Y_1 + 2(2-\nu_2)(1+\nu_2)/Y_2\right)$
Radius	$r^* = 1/\left(1/r_1 + 1/r_2\right)$
Mass	$M^* = 1/\left(1/M_1 + 1/M_2\right)$

Thus, both elastic compressibility and inelastic energy dissipation are contained. Frictional effects are included by enforcing the Coulomb criterion

$$F^{\mathrm{t}}_{j\to i} \leq \mu_{\mathrm{fr}} F^{\mathrm{n}}_{j\to i} \tag{2.5}$$

that truncates the tangential force exerted to a value relative to the normal force acting, considering the static friction coefficient μ_{fr}.

The torque due to particle contact was considered using an elastic-plastic spring pot torque model as formulated by Ai et al. (2011) that adds an elastic (superscript k) and a dissipative (superscript d) term:

$$\mathbf{T}_{j\to i} = \mathbf{T}^{\mathrm{k}}_{j\to i} + \mathbf{T}^{\mathrm{d}}_{j\to i} \tag{2.6}$$

The elastic term is given by

$$\mathbf{T}^{\mathrm{k}}_{j\to i}(t+\Delta t) = \mathbf{T}^{\mathrm{k}}_{j\to i}(t) - k_{\mathrm{rfr}}\Delta\theta \tag{2.7}$$

with k_{rfr} being the rolling stiffness and $\Delta\theta$ the *incremental* relative rotation between the particles in the shear plane. $\Delta\theta$ is limited using the coefficient of rolling friction, μ_{rfr}, relative to the normal contact force component:

$$T^{\mathrm{k}}_{j\to i} \leq \mu_{\mathrm{rfr}} r^* F^{\mathrm{n}}_{j\to i}. \tag{2.8}$$

The rolling stiffness is related to the tangential stiffness k_{t} by

$$k_{\mathrm{rfr}} = k_{\mathrm{t}}(r^*)^2. \tag{2.9}$$

In the applied variant of the elastic spring-dash pot model, the viscous damping contribution is deactivated.

The computational demand of DEM simulation lies in both the size of the time step Δt, as well as the number of elements to be considered. General performance (number of time steps per unit time) can be increased by artificially softening particles, leading to longer collision times. Lommen et al. (2014) has shown that this simplification has little influence on macroscopic system behavior. This accelerates simulations, but does not allow the consideration of smaller particles or larger industrial systems containing more than tens of millions of particles.

2.2. Governing Equations in Computational Fluid Dynamics

Fluid flow in CFD-DEM coupling is usually resolved by applying the finite volume method to the Navier–Stokes equations to yield a velocity field $\mathbf{u}_{\mathrm{f}}$ and pressure field p that are discretized on a grid consisting of cells that constitute a number of interconnected control volumes:

$$\frac{\partial(\alpha_{\mathrm{f}}\rho_{\mathrm{f}}\mathbf{u}_{\mathrm{f}})}{\partial t} + \nabla\cdot\alpha_{\mathrm{f}}\rho_{\mathrm{f}}\mathbf{u}_{\mathrm{f}}\mathbf{u}_{\mathrm{f}} = -\alpha_{\mathrm{f}}\nabla p + \alpha_{\mathrm{f}}\nabla\cdot\tau + \alpha_{\mathrm{f}}\rho_{\mathrm{f}}\mathbf{g} + \dot{\mathbf{S}}_u \tag{2.10}$$

$$\alpha_{\mathrm{f}}\frac{\partial(\alpha_{\mathrm{f}}\rho_{\mathrm{f}})}{\partial t} + \nabla\cdot\alpha_{\mathrm{f}}\rho_{\mathrm{f}}\mathbf{u}_{\mathrm{f}} = 0. \tag{2.11}$$

Here, α_{f} denotes the volumetric phase fraction of the fluid phase, ρ_{f} its density, $\mathbf{u}_{\mathrm{f}}$ its real (interstitial) velocity, $\mathbf{g}$ is the gravitational acceleration and $\dot{\mathbf{S}}_u$ is a momentum exchange term.

The pressure p is not directly solved for using the continuity equation (2.11) but is instead iterated using a pressure equation to enforce continuity. The shear stress tensor τ is given using a Newtonian law of viscosity or a more sophisticated closure. A relation between density, pressure, velocity and other state variables (most notably the temperature T_f) is required for the compressible case. Usually, the ideal gas equation $\rho_f = p/(R_f T_f)$ is used, with R_f being the mass-specific gas constant.

This formulation of the Navier-Stokes equations, referred to in literature (e.g. by Zhou et al. (2010) and Zhu et al. (2008)) as model A, includes the fluid volume fraction α_f and the momentum exchange term $\dot{\mathbf{S}}_u$ that represent the presence of a Lagrangian phase, summarized in Tab. 2.2. The other formulation, model B, considers pressure to be attributed to the fluid phase, in contrast to model A, which assumes a shared pressure among both phases. Consequently, the pressure gradient term in model B reads $-\nabla p$, and the shear stress term $\nabla \cdot \tau$. This also causes the phase interaction forces to be summed up over the total control cell volume V_{cell} in model A and the fluid volume $\alpha_f V_{\text{cell}}$ in model B, respectively.

Tab. 2.2.: Attributions of different momentum contributions in model A and model B formulations of CFD-DEM coupling.

Term	Model A	Model B
Shear stress	$\alpha_f \nabla \cdot \tau$	$\nabla \cdot \tau$
Pressure gradient	$\alpha_f \nabla p$	∇p
Momentum exchange control volume	V_{cell}	$\alpha_f V_{\text{cell}}$

The forces involved include at least

- the drag force,
- the pressure gradient force, and
- the viscous force.

In liquid-solid systems,

- the virtual mass force and
- Basset force

are required in addition to the aforementioned ones to accurately depict the effects of boundary layers (Nijssen et al., 2020).

When effects like particle rotation are to be considered,

- the Saffmann shear lift force and
- the Magnus force

contribute additional source terms that induce a torque into the fluid phase and have to be treated with further numerical effort. Although all of these forces may be present in any given physical system, their contribution might be negligible in modeling and selectively ignored to reduce computational demand.

2.2.1. Turbulence Modeling

Starting from a given flow velocity, a viscous fluid will exhibit highly non-linear, chaotic behavior due to energy dissipation in the form of vortices that form and break up on a wide range of length scales. Describing this behavior in simulations requires the spatial and temporal resolution of the range from very small length scales to large eddies in the system. Thus, the accurate representation of turbulent flows requires turbulence modeling for all practical intents and purposes. A rigorous treatment of these models can be found in the textbook by Wilcox (2006). For the fluid phase, turbulent viscosity/Reynolds-averaged Navier–Stokes models like the k-ε or k-ω models are used, which solve additional differential equations for the generation, transport, and dissipation of the turbulent kinetic energy and increase the viscosity depending on these quantities. As such, the effect of turbulence on averaged flow is captured without resolving the small length scales. Large eddy simulations, in which eddies larger than a filtering criterion are resolved, are performed less frequently. In these, the effect of those eddies occurring on smaller length scales is modeled using an isotropic model like k-ε. In both cases, the effect on particles can be stochastically modeled by using the time spent in an eddy and the turbulent velocity fluctuation, both of which depend on the turbulent kinetic energy.

2.3. Eulerian-Lagrangian Phase Coupling

As illustrated in Fig. 2.2, phase coupling is realized by introducing source-sink terms into the mass and momentum, as well as heat and species balances on the CFD side, as well as respective source-sink terms on the DEM side. The rate at which these are transferred between the phases is described using closures that normally take the fluid-mechanical and compositional situation into consideration. Algorithmically, the CFD and DEM part of the simulation are alternated, with coupling steps in between. In the coupling steps, the source-sink terms are recalculated based on the changed situation. Furthermore, mapping between the two frames of reference have to be performed, by either interpolating field quantities to the discrete elements or summing elements within the control volumes of the Eulerian mesh and creating corresponding fields.

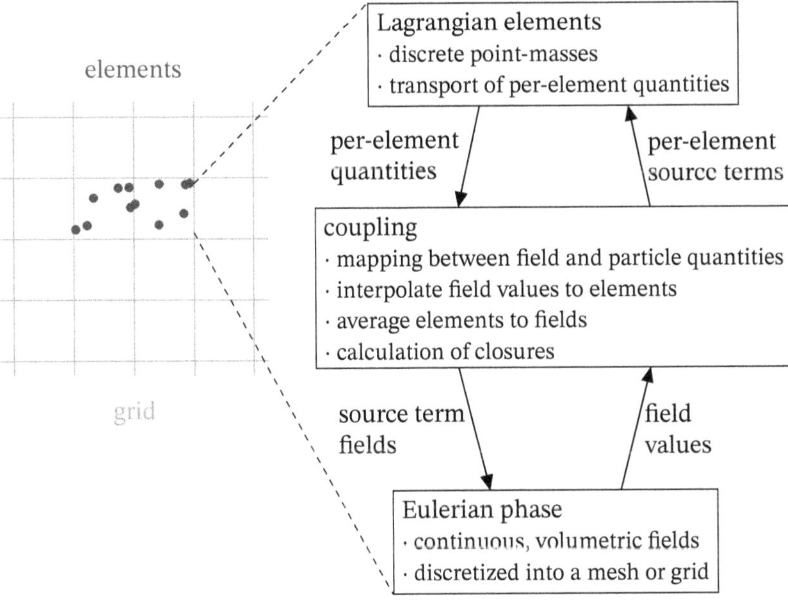

Fig. 2.2.: Schematic of Euler-Lagrange phase coupling in its most general form.

2.3.1. A Novel, Simple Coupling Algorithm for Massively Parallel Simulations

Phase coupling is implemented in two ways in the softwares used. In the OpenFOAM framework, particle or droplet motion is solved using the OpenFOAM Lagrangian Particle Tracking (OF-LPT) mechanism that moves particles through a tetrahedral decomposition of the CFD mesh. CFDEMcoupling (Goniva et al., 2012) uses the LIGGGHTS DEM code for particle motion. Most of the constituent parts of these systems appear to be mature, efficient and optimized in their parallel scaling ability, with the notable exception of the data exchange module of CFDEMcoupling-PUBLIC. Inter-process communication can be a bottleneck in parallel simulations. This section is dedicated to this topic.

Any Eulerian-Lagrangian simulation code has three components that must be considered when analyzing its parallel performance:

- choice in flow domain decomposition,
- choice in particle domain decomposition, and

- approach taken in mapping between particles and flow fields.

Unequal distribution of the workload, be it parts of the flow field to be solved or particles to be moved, will cause some processes to finish faster and wait for the others to complete, resulting in lower global efficiency.

CFDEMcoupling allows for freedom in decomposing both the flow field and particle simulation domain - a single MPI process may be responsible for spatially disparate regions when solving flow and particle motion. As such, these can be chosen for maximum efficiency and load-balancing can be used to ensure optimal distribution of particles among processes. This flexibility is paid for by additional inter-process communication taking place: in the public version of CFDEMcoupling, the communication uses an all-to-all approach, illustrated in Fig. 2.3a, where

1. memory is reserved in the coupling module to fit the entirety of particle data in each process,
2. the particle data in the local Lagrangian domain is copied into said memory and
3. *MPI_Allreduce* is called to combine the data from all domains by summing it and distribute them to all other processors.

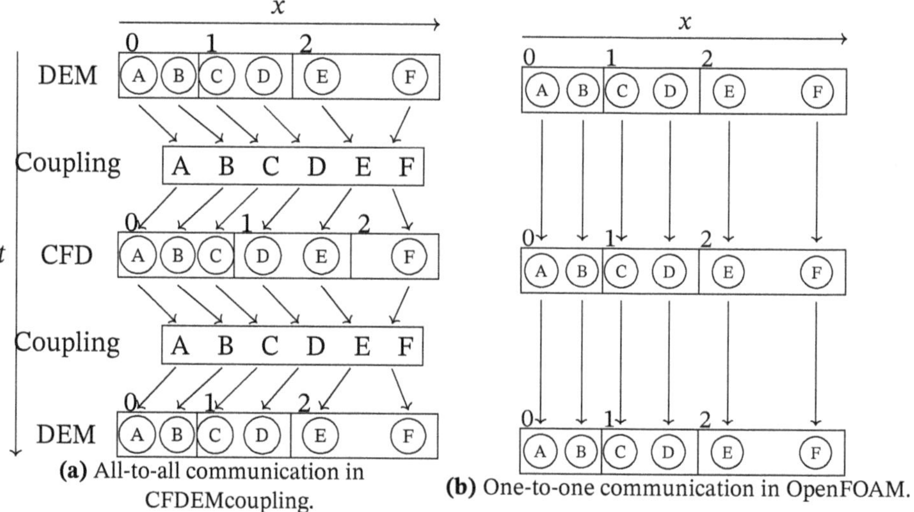

(a) All-to-all communication in CFDEMcoupling.

(b) One-to-one communication in OpenFOAM.

Fig. 2.3.: Illustration of the different communication schemes employed by CFDEMcoupling and OpenFOAM LPT. Particle identifies are labeled A through F while processors (or, equivalently, MPI ranks) are denominated by 0 to 2.

This results in excessive memory usage because data from distant domains is both received and kept in memory, as well as superfluous communication due to usage of *MPI_Allreduce*.

The resulting total memory demand and amount of data exchanged scale linearly with both the number of processes and the total number of particles. This increasing demand precludes this method from efficient use in larger simulations.

A simple one-to-one communication scheme was thus implemented to alleviate this problem while maintaining the flexibility in choice of flow and particle domains. First, the bounding boxes of individual particle and flow domains are used to establish a communication pattern that communicates all data between overlapping domains, as outlined in Fig. 2.4. In the next step of coupling, position data along with a unique identifier is transferred from DEM processes to CFD processes with matching bounding boxes, where particles are associated with cells in the actual flow domain, or recorded in a "rejection mask" if not found. For each subsequent dataset that is to be sent from the DEM processes to the CFD, all data is sent and received, but only those not contained in the mask are recorded. For reverse communication from the CFD processes to DEM, the process is performed in reverse and the identifier used to match the particle data to its origin and position in memory. The memory required in each process and amount of data communicated is greatly reduced - although the decomposition choices will affect the number of other processes that have to communicate.

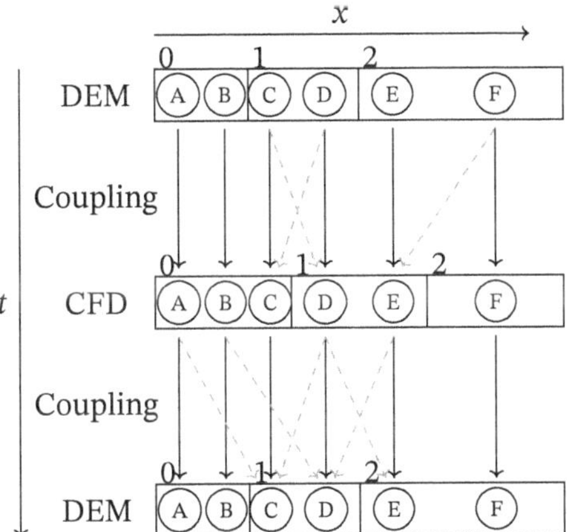

Fig. 2.4.: Proposed one-to-one communication scheme for CFDEMcoupling. Particle identifies are labeled A through F while processors (or, equivalently, MPI ranks) are denominated by 0 to 2.

OpenFOAM LPT follows a fundamentally different approach. It tracks particle motion using local, barycentric coordinates relative to a tetrahedral decomposition of the CFD mesh, meaning that flow and particle domains are identical (see Fig. 2.3b). This allows for close coupling of flow and single particle models, as the numerical expense of mapping is minimal and no separate coupling step besides the construction of source terms for the Eulerian equations is present. The flexibility in choosing the optimal decomposition for either particle motion or solving the flow equation is reduced, but the system has constant total memory demand and only particles crossing between neighboring domains are communicated. This has to be weighed with the numerical expense of tracking particles.

2.3.2. Euler-Lagrange Mapping

Due to the disparate natures of the Lagrangian elements (particles/droplets) and the flow fields stored in an Eulerian frame of reference, a mapping between these has to be established. The issue of solids-phase volume fraction calculation and mapping of properties from the discrete to the continuous phase is approached in several ways. The most popular is the iteration over all particles present in the cell and averaging/summing up their respective values, for example, the explicit momentum source term from particles to fluid:

$$\dot{\mathbf{S}}_{u,j} = \frac{\sum_{\text{particles } i \text{ in cell } j} \mathbf{F}_{\text{drag},i}}{V_{\text{cell},j}} \tag{2.12}$$

This very simplistic approach has the advantage of being very numerically efficient but introduces issues with fine grid cells and inaccuracies at cell boundaries. Thus, a decomposition of the particles into a cloud of N subpoints with a numerical weight $\phi_{i,j}$ may be performed:

$$\dot{\mathbf{S}}_{u,j} = \frac{\sum_{\text{particles } i \text{ close to cell } j} \phi_{i,j} \mathbf{F}_{\text{drag},i}}{V_{\text{cell},j}} \tag{2.13}$$

With the subpoints for each particle i summing to unity over all cells that are touched by the subpoints:

$$\sum_{\text{cells } j} \phi_{i,j} = 1 \tag{2.14}$$

The added computational demand lies in associating the subpoints with grid cells — a process that is somewhat expensive in unstructured grids. The choice in the position of sub points, as well as the choice in weighting, leads to a variety of options for the end user or developer. The issue in using the center point method is illustrated in Fig. 2.5. When particles are large relative to the grid, the source term can be localized in the cell in which the center point is located rather than distributed over subpoints that are located in other cells the particles partially covers.

The problem of numerical instability owing to particles covering entire cells or very large source terms may also be addressed via a diffusion approach introduced by Pirker et al. (2011), in which the corresponding source terms are spread out using an implicit diffusion equation.

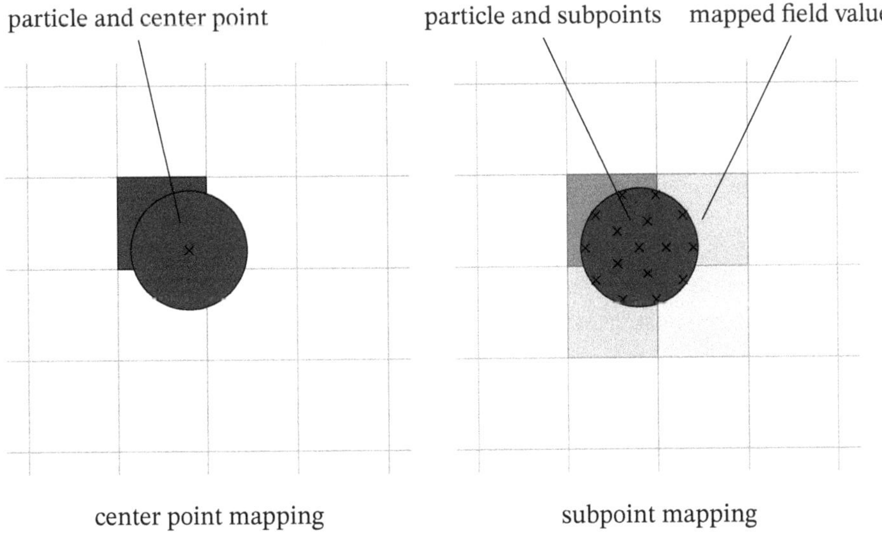

Fig. 2.5.: Schematic of the consequence of using the center point mapping (left) and the sub point mapping (right).

2.3.3. Momentum Coupling for Dense Particulate Flows

The previous sections explained the motion of the particle phase is governed by the equations of motion and the fluid motion is solved on a grid. What remains is the rate at which momentum is exchanged, namely the values that

- the drag force $\mathbf{F}_{i,\mathrm{drag}}$ that a given particle i experiences and
- the momentum exchange term $\dot{\mathbf{S}}_\mathrm{u}$

take. In the unresolved CFD-DEM method, where particles are much smaller than flow cells, this is modeled using the drag laws and other closures summarized in section 2.2. These provide a closure for the drag force

$$\mathbf{F}_{i,\mathrm{drag}} = \frac{\pi}{6}\frac{\beta_i}{\alpha_\mathrm{p}}\left(\mathbf{u}_\mathrm{f} - \mathbf{u}_{i,\mathrm{p}}\right) \tag{2.15}$$

where β_i is the momentum exchange coefficient that correlations describe, depending of the complex hydrodynamics happening in relation to the particle Reynolds number

$$Re_\mathrm{p} = \frac{||\mathbf{u}_\mathrm{f} - \mathbf{u}_\mathrm{p}||\alpha_\mathrm{f} d_\mathrm{p}}{\eta_\mathrm{f}} \tag{2.16}$$

and the particle solids volume fraction α_p in the vincinity of the particle. The key material property of the fluid phase in this regard is the viscosity η_f. The most popular correlation is that of Gidaspow (1994), which uses the empirical correlation of Ergun and Orning (1949), as formulated by Beetstra et al. (2007),

$$\beta = 150 \frac{\eta_f}{d_p^2} \frac{\alpha_p^2}{(1-\alpha_p)} + 1.75 \frac{\eta_f}{d_p^2} \frac{\alpha_p}{1-\alpha_p} Re_p \tag{2.17}$$

for flow below the minimum fluidization velocity and switches to the correlation of Wen & Yu (16, 17) for higher gas velocities:

$$\beta = \begin{cases} \frac{18\eta_f \alpha_p (1-\alpha_p)(1-\alpha_p)^{-3.65}(1+Re_p^{0.687})}{d_p^2} & Re_p < 1000 \\ \frac{0.33\eta_f \alpha_p (1-\alpha_p)(1-\alpha_p)^{-3.65}}{d_p^2} & Re_p \geq 1000. \end{cases} \tag{2.18}$$

Both underlying relationships were derived from experimental investigations that have their uncertainties. Hill et al. (2001), Beetstra et al. (2007) and Tenneti et al. (2011), among many others, have applied the lattice Boltzmann and immersed boundary CFD-DEM methods to successively more complex particle flow scenarios to derive correlations between drag force and flow conditions—at first only for static arrays of spheres for low Reynolds numbers, then increasing the range of Reynolds numbers, and finally allowing the particles to move. Among these, the correlation according to Beetstra et al. (2007) is particularly popular:

$$\beta = \pi d_p \nu_f \alpha_f \rho_f F, \tag{2.19}$$

$$F = \frac{10\alpha_p}{\alpha_f^2} + \alpha_f^2 \left(1 + 1.5\alpha_f^{1/2}\right) \frac{0.413 Re_p}{24\alpha_f^2} \left(\frac{\alpha_f^{-1} + 3\alpha_f \alpha_p + 8.4 Re_p^{-0.343}}{1 + 10^{3\alpha_p} Re^{-0.5-2\alpha_p}} \right). \tag{2.20}$$

This correlation is used in all of the CFD-DEM simulations performed in this thesis.

In its simplest terms, the source term for the fluid phase is given by

$$\dot{\mathbf{S}}_u = \beta \frac{\alpha_f}{1-\alpha_f} \mathbf{u}_f - \beta \frac{\alpha_f}{1-\alpha_f} \bar{\mathbf{u}}_p \tag{2.21}$$

where $\bar{\mathbf{u}}_p$ is the averaged particle velocity. The effect of displacement of fluid by the solid phase is considered in the form of the volumetric phase fraction of the fluid phase

$$\alpha_f = \frac{\sum_j \phi_j V_{p,j}}{V_{cell}}, \tag{2.22}$$

where $V_{p,j}$ is the the particle volume and ϕ_j is the per-particle contribution fraction.

The general flow of information for coupling is shown in Fig. 2.6.

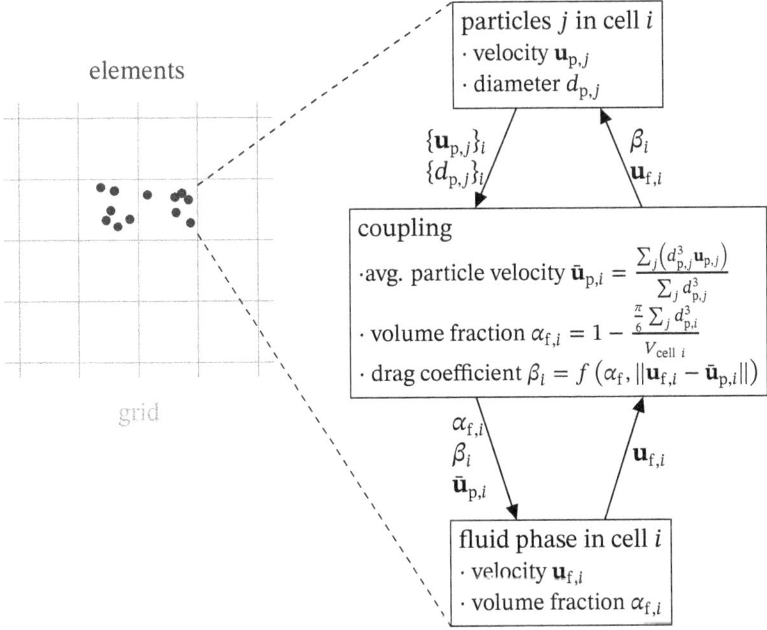

Fig. 2.6.: Schematic of the flow of state information in coupling between particle and fluid phase in a coupled CFD-DEM simulation.

During time integration, calculation is periodically interrupted to recalculate the source terms. To this end, mapping and averaging is performed for the relevant quantities such as velocities and the volume fraction. The drag coefficient β_i is recalculated and communicated to both the fluid and particle phase. For the evaluation of the drag force acting upon particles, the fluid velocity is communicated as well, so that Eq. 2.15 can be evaluated semi-implicitly as in the given form. Analogously, the fluid velocity-dependent first term of the fluid phase momentum source term in Eq. 2.21 is integrated into the left-hand-side of the system equations and, thus, solved implicitly. The fixed last term is integrated into the right-hand-side. In using this semi-implicit procedure, strongly coupled flows can be simulated with large time steps while retaining stability and accuracy (Radl et al., 2014). The peculiarities of the implementation used in CFDEMcoupling can be found in the works by Goniva et al. (2012).

2.4. Coarse-Graining Techniques for Dense Particulate Flows

Coarse-graining techniques aim to lower computational cost by reducing the overall spatial/temporal resolution of the fluid part and/or the particle part of the simulation while preserving macroscopic quantities and the overall temporal evolution of the system.

2.4.1. Particle Coarsening

Owing to the computational cost of modeling the collision dynamics of every single grain in a granular system, particle coarsening

- reduces the number of particles that must be tracked, so larger apparatuses and/or finer particles can be simulated, and
- extends the process time that can be simulated by permitting a larger time step.

This makes particle coarsening nearly unavoidable and highly beneficial when dealing with industrial-scale apparatuses. In the literature, such collectives of particles with identical properties are referred to as parcels. Most commonly, particles are enlarged by a factor of δ_{CG}. This decreases the number of particles by a factor of δ_{CG}^3 and increases the permittable time step by a factor of approximately δ_{CG}, as a rule of thumb. The DEM contact model coefficients are modified to preserve the bulk behavior by either using an *a priori* approach, e.g., by using the approach of Bierwisch et al. (2009) or Benyahia and Galvin (2010), or calibrating in a way that allows the macroscopic behavior of fine, real particles to be represented with larger numerical parcels (Kieckhefen et al., 2018b). Particle coarsening requires physical models to base their calculations on single-particle values (e.g., surface area) and then scale source terms with a factor δ_{CG}^3. This ensures that the number of primary particles per parcel are considered either according to Radl et al. (2011), or in other manners that are appropriate to the exact coarsening criteria. For example, specific coarsening criteria can be chosen for use with liquid–solid systems (Lu et al., 2016). The use of particle coarsening techniques introduces the need to coarsen the CFD or coupling grid, and to allow for accurate mapping and applicable phase fraction calculations.

2.4.2. Fluid Coarsening

Coarser grids induce loss of local structure representation, as illustrated in Fig. 2.7. The loss of sharp boundaries, for instance, of the edges of particle clusters, induces the overestimation of drag force on particles.

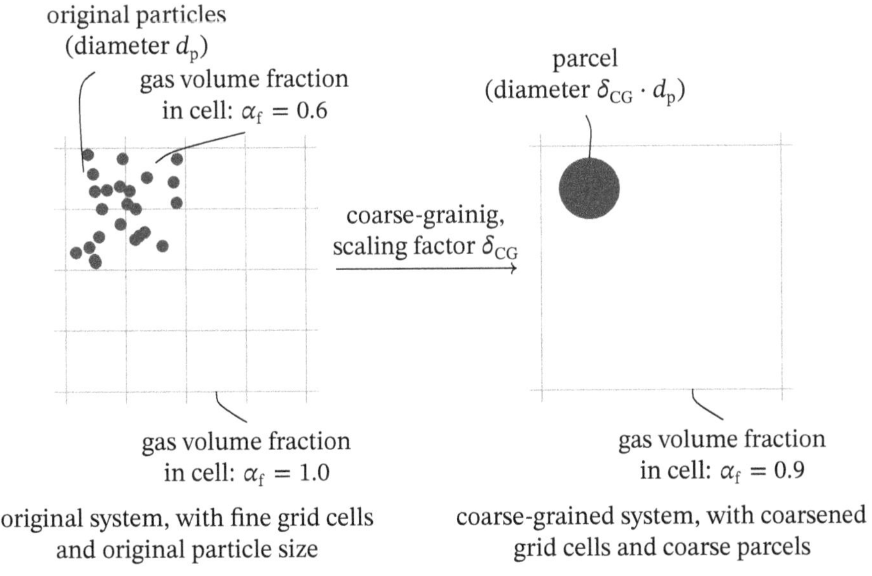

Fig. 2.7.: Schematic of the issue of loss of structure when using particle and/or fluid coarsening techniques in strongly coupled granular flows.

To correct for this, drag law corrections have been developed by performing simulations in which particles can swarm in a periodic box with gravity and a preset pressure differential for various Froude numbers and solids volume fractions on both refined and coarse grids (Ozel et al., 2016). A different class of drag models named energy-minimization multiscale models (Li and Kwauk, 1994), aims to tackle this problem by correlating the fraction of particles that will cluster up to a certain size in each cell with the flow situation and, possibly, geometric parameters (i.e., distance to a wall) to predict the net drag force that will act on the particles. This is of interest in the context of industrial-scale fluidized beds, where applications of both fluid and particle coarsening are inevitable. An alternative to these kinds of models are the filtered drag models (Ozel et al., 2016; Schneiderbauer and Pirker, 2014). These offer a correction to regular drag laws and are obtained by performing simulations of particles falling in a fluid on successively coarser grids for different flow conditions and solids-phase fractions.

3 Development of a Segregated Calibration Approach for Handling Wetted Materials in the Discrete Element Method

The governing equations defined in section 2.1 can be used to extend the DEM to consider liquid-containing systems-most prominently, the effect that liquids have on granular dynamics. To this end, a short overview of the micromechanics, with a focus on the implementation in the context of DEM simulations, is given. A novel closure for wetted contact probability and liquid involved in modelling partially wetted systems using DEM codes is developed and presented. The sensitivity of liquid content in the system and the pertinent model parameters with respect to a set of bulk solid characterization experiments are assessed. Based on this, a novel, segregated calibration approach is proposed.

3.1. Micromechanics of Wetted Granular Matter

In solids, the presence of liquids tends to leads to increased dissipation and a decrease in flowability. The contact amongst wetted granules, wetted granules with walls or dry particles, as well as the presence of surface liquid introduces attractive forces in addition to the commonly considered contact forces. Buck et al. (2017) distinguished four main phases:

1. *fluid displacement*: When the granules in question approach each other and the liquid films touch, a collision occurs. As the collision continues, the liquid film is displaced. This phase is dominated by viscous forces, as the relative velocity is still high and there is no liquid bridge beyond the original liquid film height.
2. *collision*: Solids surfaces touch and contact forces dominate.
3. *retraction and liquid bridge stretching*: Both viscous and capillary forces apply due to the dissipation and reversal of the velocity vector w.r.t. the normal direction. As the liquid bridge continues stretching, the capillary force may start to substantially contribute to energy dissipation.

4. *liquid bridge rupture*: At the critical point, the liquid bridge ruptures and the liquid contained in the bridge returns to the granules. At this point, the viscous and capillary force cease to apply.

The dominant component in the process is dependent on the liquid and collision velocity. The capillary number

$$Ca = \frac{\eta_l u_r}{\sigma_{surface}} \tag{3.1}$$

is a common measure of the ratio of viscous to surface tension forces present during such collisions. The pertinent material property for the surface tension force is the surface tension $\sigma_{surface}$ and the major factors for the viscous force are the viscosity η_l and the collision velocity u_r. In situations with low relative velocities, weakly viscous liquids, or high surface tension, the capillary number will be high and capillary forces will be much more significant than viscous forces. Such situations can be found in moving beds, granular mixers or rotating drums with low velocities, where granular shear flows apply. In contrast, high velocities, viscosities and low surface tensions result in high capillary numbers, necessitating the consideration of viscous forces. Such situations are present in fluidized/spouted beds, high shear granulators and solid mixers with high energy input.

3.1.1. Capillary Force

Capillary forces occur due to the geometry of the solid surface and the surface tension of the liquid on the solid surface. The most general description of the physics involved is the Young-Laplace equation that describes the normal stress of a surface at rest

$$\Delta p_{cap} = -\sigma_l(\nabla \cdot \mathbf{e}_n), \tag{3.2}$$

where Δp_{cap} is the pressure jump across the surface, σ_l is the surface tension and $\mathbf{e}_n$ is the normal vector of the surface, as illustrated in Fig. 3.1.

The term $\nabla \cdot \mathbf{e}_n$ is thereby the curvature of the surface. The higher the pressure jump, the larger the driving force of the system to reduce the surface area, and correspondingly the curvature of the surface. Lian et al. (1993) describe an exact solution for the force equilibrium between two spheres of radius r_p, given by summing two contributions:

$$\mathbf{F}_{cap} = \left(\underbrace{\sigma_{surface} \sin(\varphi + \theta) \overbrace{(2\pi r_p \sin(\varphi))}^{\text{contact line}}}_{\text{due to three-phase contact}} + \underbrace{\overbrace{\pi(\sin(\varphi) r_p)^2}^{\text{interface area}} \Delta p_{cap}}_{\text{due to underpressure in bridge}} \right) \mathbf{e}_n. \tag{3.3}$$

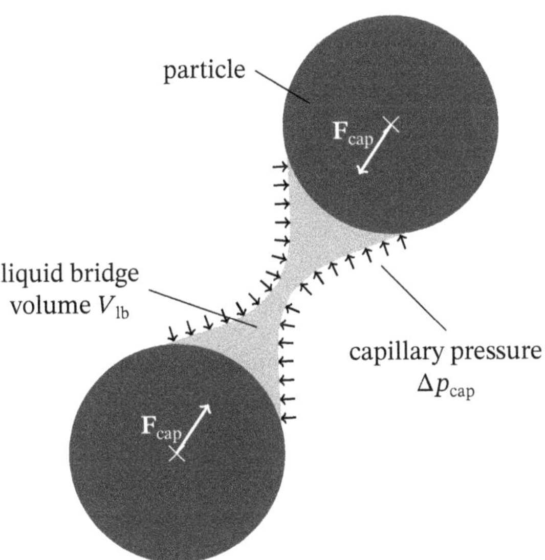

Fig. 3.1.: Schematic of a liquid bridge with the capillary pressure curving the bridge and resulting in the net capillary force acting upon the particles. The arrows indicate the normal direction of the capillary bridge surface.

This closed expression still depends on the half-filling angle φ, defined as the angle between the axis of the liquid bridge and the outermost liquid attachment points, and the three-phase contact angle θ. A closure for the pressure drop Δp_{cap} is often derived by approximating a shape of the bridge. The first contribution is due to the three-phase contact line at the particle surface. The term $2\pi r \sin(\varphi))$ is the circumference of this contact line, and the factor $\sin(\varphi + \theta)$ considers that the force is exclusively active in the normal direction - the tangential component cancels out under the assumption of axial symmetry. The second contribution is the force that acts in the normal direction due to the capillary pressure. The capillary underpressure that minimizes the surface of the bridge by curving it, also applies to the particle-liquid bridge area $\pi(r_{\text{p}} \sin(\varphi))^2$. This formulation introduces issues when applied in DEM, namely:

- *open-form expression*: By not providing an explicit closure for Δp_{cap}, the implementation as a force model is made impossible without further solvers for the Young-Laplace equation (3.2).
- *lack of relation to particle quantities*: When applying DEM, the state of a particle has to be clearly defined by a number of scalar values that are tracked for every particle, for example the liquid volume available at the particle surface.

Mikami et al. (1998) nondimensionalized (3.2) and transformed it to a cylindrical coordinate system. The resulting equations were solved numerically for a variety of liquid bridge volumes and contact angles. Consequently, an explicit capillary force law was derived:

$$\mathbf{F}_{\mathrm{cap}} = \pi r_{\mathrm{p}} \sigma_{\mathrm{l}} \left(\exp\left(A \frac{\delta_{\mathrm{lb}}}{r_{\mathrm{p}}} + B \right) + C \right), \tag{3.4}$$

in which parameters A and C depend solely on the nondimensional liquid bridge volume $V_{\mathrm{lb,red}} = V_{\mathrm{lb}}/r_{\mathrm{p}}^3$, and B on the contact angle, in addition to the former. C. Boyce (2018) compared the results of the capillary force law of Mikami et al. (1998) to Willett et al. (2000) and found good agreement.

Experimentally, the capillary force in liquid bridges are either studied using restitution experiments or atomic force microscopy (AFM) experiments. In AFM, wetted spheres are brought into contact and the relationship between bridge geometry, distance and resulting force is studied. One such study was conducted by Rabinovich et al. (2005), who applied atomic force microscopy with an accuracy of 18.5% to glass spheres of 20 µm-50 µm and silicon surfaces, wetted with oil. Their experiments confirm the previously theoretically-derived expressions which are give in Tab. 3.1. The assumption that the dynamic change of the contact angle can be neglected was verified.

Tab. 3.1.: Capillary force model of Rabinovich et al. (2005).

quantity	expression	
capillary force		
$(\delta_{\mathrm{lb}} > 0)$	$\mathbf{F}_{\mathrm{cap}} = -2\pi r_{\mathrm{eff}} \sigma_{\mathrm{surface}} \left(\frac{\cos(\theta_{\mathrm{eff}})}{1+\delta_{\mathrm{lb}}/(2d_{\mathrm{rab,p\text{-}p}})} + \sin(\theta_{\mathrm{rab}}) \sin(\theta + \theta_{\mathrm{eff}}) \right) \mathbf{e}_{\mathrm{n}}$	(3.5)
$(\delta_{\mathrm{lb}} = 0)$	$\mathbf{F}_{\mathrm{cap}} = -2\pi r_{\mathrm{eff}} \sigma_{\mathrm{surface}} \left(1 - \delta_{\mathrm{lb}} \sqrt{\frac{\pi r_{\mathrm{eff}}}{V_{\mathrm{lb}}}} \right) \mathbf{e}_{\mathrm{n}}$	(3.6)
	$\theta_{\mathrm{rab}} = \sqrt{\gamma_{\mathrm{rab}} \frac{\delta_{\mathrm{lb}}}{r_{\mathrm{eff}}}}$	(3.7)
model parameters	$d_{\mathrm{rab,p\text{-}p}} = \frac{1}{2} \delta_{\mathrm{lb}} \gamma_{\mathrm{rab}}$	(3.8)
	$\gamma_{\mathrm{rab}} = -1 + \sqrt{1 + \frac{2V_{\mathrm{lb}}}{\pi r_{\mathrm{eff}} \delta_{\mathrm{lb}}^2}}$	(3.9)

The proposed capillary force law contains no fitted factors and, like the law of Mikami et al. (1998), depends only on distance, bridge volume and particle radii. For the case of particle surface contact, the derived law takes a different form: the term considering the capillary pressure goes to a fixed value as the separation approaches zero.

However, Lambert et al. (2008) found the contribution of the three-phase contact term $-2\pi r_{\mathrm{eff}} \sigma_{\mathrm{surface}} \sin(\theta_{\mathrm{rab}}) \sin(\theta + \theta_{\mathrm{eff}})$ to be detrimental for larger volumes. As such, they recommend the use of

$$\mathbf{F}_{\text{cap}} = -2\pi r_{\text{eff}}\sigma_{\text{surface}}\frac{\cos(\theta_{\text{eff}})}{1+\delta_{\text{lb}}/(2d_{\text{rab,p-p}})}\mathbf{e}_{\text{n}}. \tag{3.10}$$

Soulié et al. (2006) extended the work by Mikami et al. (1998) by additionally considering polydispersity. The model formulation is given in Tab. 3.2.

Tab. 3.2.: Capillary force model of Soulié et al. (2006).

quantity	expression	
capillary force	$\mathbf{F}_{\text{cap}} = -\pi\sigma_{\text{surface}}\sqrt{r_i r_j}\left(\exp\left(A_{\text{soulie}}\frac{\delta_{\text{lb}}}{\max(r_i,r_j)} + B_{\text{soulie}}\right) + C_{\text{soulie}}\right)\mathbf{e}_{\text{n}}$	(3.11)
reduced bridge volume	$V_{\text{lb,red}} = \frac{V_{\text{lb}}}{(\max(r_i,r_j))^3}$	(3.12)
model parameters	$A_{\text{soulie}} = -1.1V_{\text{lb,red}}^{-0.53}$	(3.13)
	$B_{\text{soulie}} = (-0.148\log(V_{\text{lb,red}}) - 0.96)\theta_{\text{eff}}^2 - 0.0082\log(V_{\text{lb,red}}) + 0.48$	(3.14)
	$C_{\text{soulie}} = 0.0018\log(V_{\text{lb,red}}) + 0.078$	(3.15)

The radius of the Mikami expression is replaced by the geometric mean $\sqrt{r_i r_j}$ and the liquid bridge volume is normalized relative to the radius of the larger particle. While this relation is still not perfect - it performed worse in their validation using steel spheres for larger diameter ratios - it is still able to capture the general shape of the force-distance curve, especially for larger distances.

A comparison between the capillary force given by the Rabinovich et al. (2005) (Tab. 3.1) and that by Soulié et al. (2006) (Tab. 3.2) for a constant surface tension and contact angle is given in Fig. 3.2. While both models predict the similar forces for identical separations and bridge volumes for small bridge volumes $V_{\text{lb,red}} < 1 \cdot 10^{-5}$ - largely independent of the volume of the liquid bridge - the correlation of Soulié et al. (2006) predicts an asymptotic independence of the separation for large bridge volumes, with the asymptote being the force acting when the particles are in contact. The Rabinovich et al. (2005) model, too, predicts an asymptotic independence of the separation for large bridge volumes, albeit with the capillary force becoming much stronger with higher liquid bridge volumes.

Antonyuk et al. (2009) identified the role of surface roughness as a minimum distance parameter for the capillary force as a fitting parameter to reproduce viscous dissipation. Capillary forces cause a dissipative effect on particle contacts. This is partially because they accelerate particles that approach each other, despite being symmetric potentials.

This, in turn, increases the viscous forces cause the dissipation, during the action of both contact forces and liquid viscous forces. The capillary and viscous forces interact to create

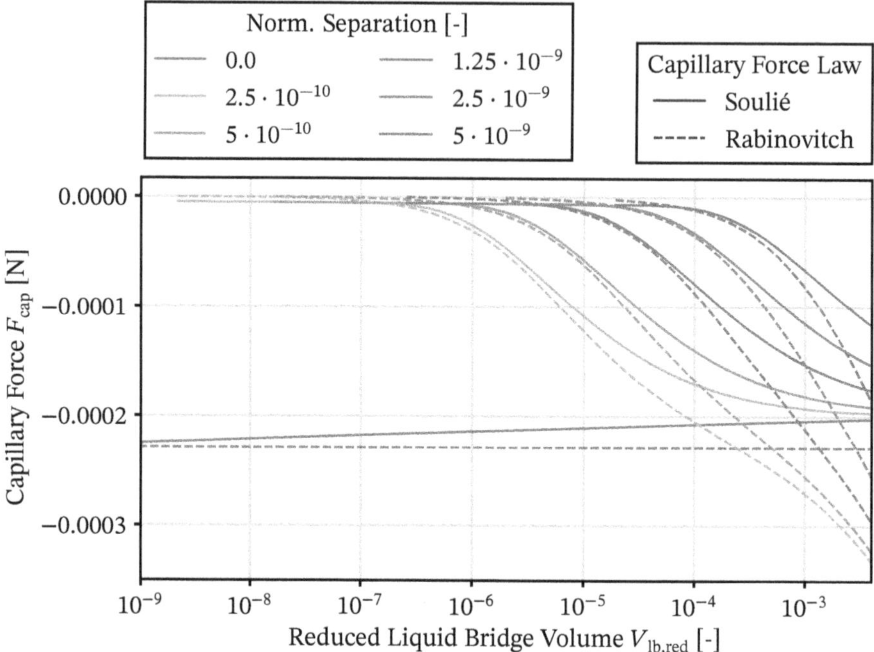

Fig. 3.2.: Capillary force responses for the Rabinovitch (3.5) and Soulié (3.12) equations, given as a function of the reduced liquid bridge volume for a contact angle $\theta = 0$ and a surface tension $\sigma_{\text{surface}} = 72.8\,\text{mN}\,\text{m}^{-1}$.

a positive feedback loop, increasing the dissipation (see section 3.1.2). Additionally, the asymmetry between liquid layer contact time during approach and reproach - due to liquid bridge stretching - contributes to a lesser degree.

3.1.2. Viscous Force

The viscous force component of a particle-particle or particle-wall collision describes the force that acts due to displacement of surface liquid. Nase et al. (2001) used an expression from lubrication theory that contains no dependency on liquid bridge volume:

$$\mathbf{F}_{\text{visc}} = -6\pi\eta_{\text{l}} \left(\mathbf{u}_{\text{p,n}} \frac{r_{\text{eff}}}{\max\left(\delta_{\text{lb,min}}, \delta\right)} + \mathbf{u}_{\text{p,t,surface}} \left(\frac{8}{15} \log\left(\frac{r_{\text{eff}}}{\max\left(\delta_{\text{lb,min}}, \delta\right)} \right) + 0.9588 \right) \right) \tag{3.16}$$

As previously elaborated upon by Antonyuk et al. (2009), the reciprocal dependence on the particle surface distance, the minimum value of the surface separation is used in case of low particle separations due to the possibility of overlap in discrete element simulations. This can be seen in detail in Fig. 3.3, where the normalized viscous force is plotted against the separation for different velocities. As the separation δ_{lb} approaches zero, the force grows with $F_{\mathrm{visc}} \propto \delta_{\mathrm{lb}}^{-1}$. Choosing the cut-off level at a lower distance increases the dissipative effect because a stronger viscous force that, by definition, acts counter to the direction of motion. Thus, the choice of $\delta_{\mathrm{lb,min}}$ is of great importance in modulating the velocity-dependence of wetted contacts.

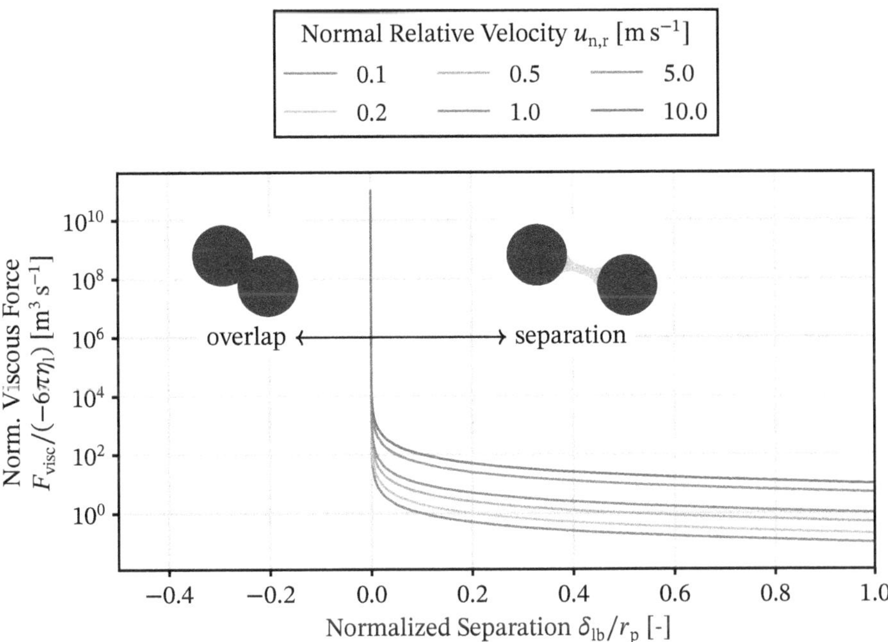

Fig. 3.3.: Normal viscous force according Nase et al. (2001) for different velocities and particle separations.

3.1.3. Liquid Bridge Rupture and Liquid Distribution

The liquid bridge rupture criterion is important because it determines the point at which the capillary force ceases to act. To this end, Lian et al. (1993) determined an expression for the free surface energy of a liquid bridge in terms of the separation and liquid bridge volume that has both a stable and an unstable solution. The meeting point of these solutions yields a closed-form expression for liquid bridge rupture that solely depends on the volume of the liquid bridge:

$$\delta_{\text{lb,rupture}} = (1 + 0.5\theta)\sqrt[3]{V_{\text{lb}}} \tag{3.17}$$

The redistribution of liquid can be done by a proportional scaling approach:

$$\Delta M_{\text{l,lb}\to i} = \rho_{\text{l}} V_{\text{lb}} \frac{r_i^n}{r_i^n + r_j^n}, \tag{3.18}$$

where n is an exponent representing the dimensionality of the proportional split. $n = 1$ would be a particle diameter-proportional split, $n = 2$ an particle surface area-proportional split and $n = 3$ a particle volume-proportional split.

3.1.4. Liquid Bridge Volume

Wetting has an influence on particle systems at very low liquid loadings. In the somewhat extreme example of water and glass beads of a size of $d_{\text{p},i} = 1\,\text{mm}$, cohesive effects become noticeable at water loadings of $X_{\text{p}} = 0.2\,\text{g}\,\text{kg}^{-1}$. Here, pouring particles from a height of about 10 cm onto a tabletop does not result in immediate dispersal due to their high coefficient of restitution and low rolling friction. Starting with $X_{\text{p}} = 0.4\,\text{g}\,\text{kg}^{-1}$, their pouring behavior is much more gel-like: they form clumps that can be readily deformed. The behavior is fully plastic in that there is resistance to macroscopic deformation but no restoring force once the deformation is applied.

A crux in the development of discrete element models of wetted systems in the pendular regime is the lack of models for surface wetting state. All categories of force models for liquid bridge forces, be they in the form of explicit force models or a mere reduction of the coefficient of restitution, require a closure for the amount of liquid V_{lb} of the overall amount of surface liquid $V_{\text{l},i} = M_{\text{l},i}/\rho_{\text{l}}$ that participates in a liquid bond. Fundamental research has been carried out by Buck et al. (2018) by capturing the amount of liquid in a liquid bridge that is present using high-speed camera recordings of the impact of dry particles on a wetted plate. These values were used in conjunction with a set of liquid bridge force and Hertzian contact models to accurately represent the resulting coefficient of restitution. This lacks predictiveness and applicability, as these do not cover the range of very low layer heights commonly encountered. The inhomogeneity of liquid on a particle surface and the consequence of this for the formation of liquid bonds is, at most, treated in surface-discretized simulations as those carried out by Schmelzle et al. (2018), at tremendous computational cost. Inhomogeneous surface liquid distributions are characterized by the property that not every contact contributes liquid for a liquid bridge. The assumption of homogeneous surface liquid distribution also diminishes the strength of other contacts, as a spotty liquid distribution would increase the volume of liquid bridges that are established. The assumption of that either all of the liquid or a fraction gets used in a liquid bond, at every single contact leads to a misrepresentation of the systems behavior, as illustrated in Fig. 3.4.

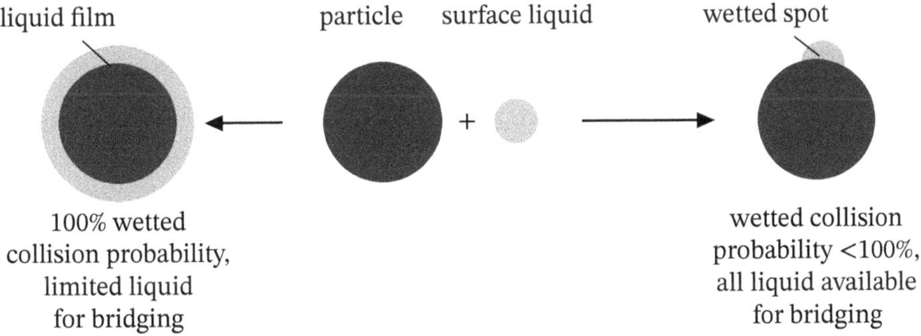

Fig. 3.4.: Previous state of the art: choosing between low layer heights and high contact probability, or high liquid layer heights and low probability.

Thus, a model is proposed that uses the two model quantities,

- contact probability ξ and
- participating liquid fraction ψ,

along with regime-change thresholds as fitting parameters to yield an explicit mapping $V_{\mathrm{lb},i\rightarrow\mathrm{lb}} = f\left(V_{\mathrm{l},i}\right)$ between surface liquid content and liquid bond volume. Furthermore, an estimate for the surface liquid layer height h_i has to be given for the contact detection. It can be put into the general form

$$\Delta V_{i\rightarrow\mathrm{lb}} = N(\xi_i)\psi_i V_{\mathrm{l},i}, \tag{3.19}$$

where $N(\xi)$ is 1 with a probability of ξ and 0 with a probability of $(1-\xi)$.

Shi and McCarthy (2008) give the maximal fraction of surface of a particle that can contribute to a liquid bridge as:

$$\psi_{\max,i} = \frac{1 - \frac{r_i}{\sqrt{r_i^2 + r_j^2}}}{2}. \tag{3.20}$$

This is taken as the maximum surface fraction of liquid that can be contributed to a liquid bridge.

Three different regimes are differentiated to blend between the two edge cases represented in Fig. 3.4:

1. *sparse coverage*: If a particle's surface liquid content is below a given threshold ($V_{s,i} < V_{thr,1}$, where $\varphi_{cov,i}\left(V_{thr,1}\right) = \psi_{max,i}$), all of its surface liquid has to participate in the contact ($\psi = 1$). The probability of having any contribution $\xi = \varphi_{cov}$ is estimated by the surface coverage fraction $\varphi = \frac{3V_{l,i}}{4\pi\left(\left(h_{min}+r_i\right)^3 - r_i^3\right)}$ at the minimum layer height $h = h_{min}$.

2. *continuous coverage*: If the surface liquid content is above this first threshold, but below the second threshold ($V_{thr,1} < V_{s,i} < V_{thr,2}$, where $\varphi_{cov}(V_{thr,2}) = 1$, we assume that the contact probability is the same as before ($\xi_i = \varphi_{cov,i}$), but the participating surface liquid fraction is given by $\psi_i = \psi_{max,i} \frac{V_{thr,2}}{V_{l,i}}$ so that the liquid layer height is $h_i = h_{min}$.

3. *full coverage*: If the surface liquid content exceeds the second threshold ($V_{s,i} > V_{thr,2}$), the contact probability is $\xi_i = 1$, and the participating liquid fraction ($\psi_i = \psi_{max,i}$) is calculated using (3.20). The layer height depends on the amount of liquid present on the surface and is given by $h = \sqrt[3]{r_i^3 + \frac{3V_{l,i}}{4\pi}} - r_i$.

The two thresholds are defined to ensure continuity among the regimes. The threshold between *full* and *continuous coverage* is trivially given by

$$V_{thr,2} = \frac{4\pi}{3}\left(\left(h_{min} + r_i\right)^3 - r_i^3\right). \tag{3.21}$$

The threshold is chosen so that the volume is conserved under the assumption of $h_i = h_{min}$:

$$V_{thr,1} = \psi_{max,i} V_{thr,2}. \tag{3.22}$$

This leaves this model with one key parameter: the minimum surface liquid layer height h_{min} that affects both contact probability and regime transitions. These could theoretically be decoupled to gain more degrees of freedom.

The general characteristics are schematically illustrated in Fig. 3.5 and the state quantities plotted for an example case in 3.6.

As intended, the probability of contributing liquid to the bridge is a stochastic process. In the sparse regime, the probability of having a nonzero contribution is low but increases steadily throughout the sparse and continuous regime until full coverage is achieved.

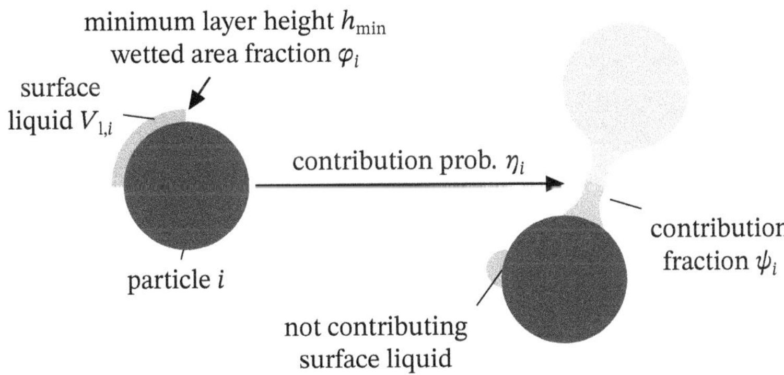

Fig. 3.5.: Illustration of the scenario the novel liquid distribution model can describe using a single parameter.

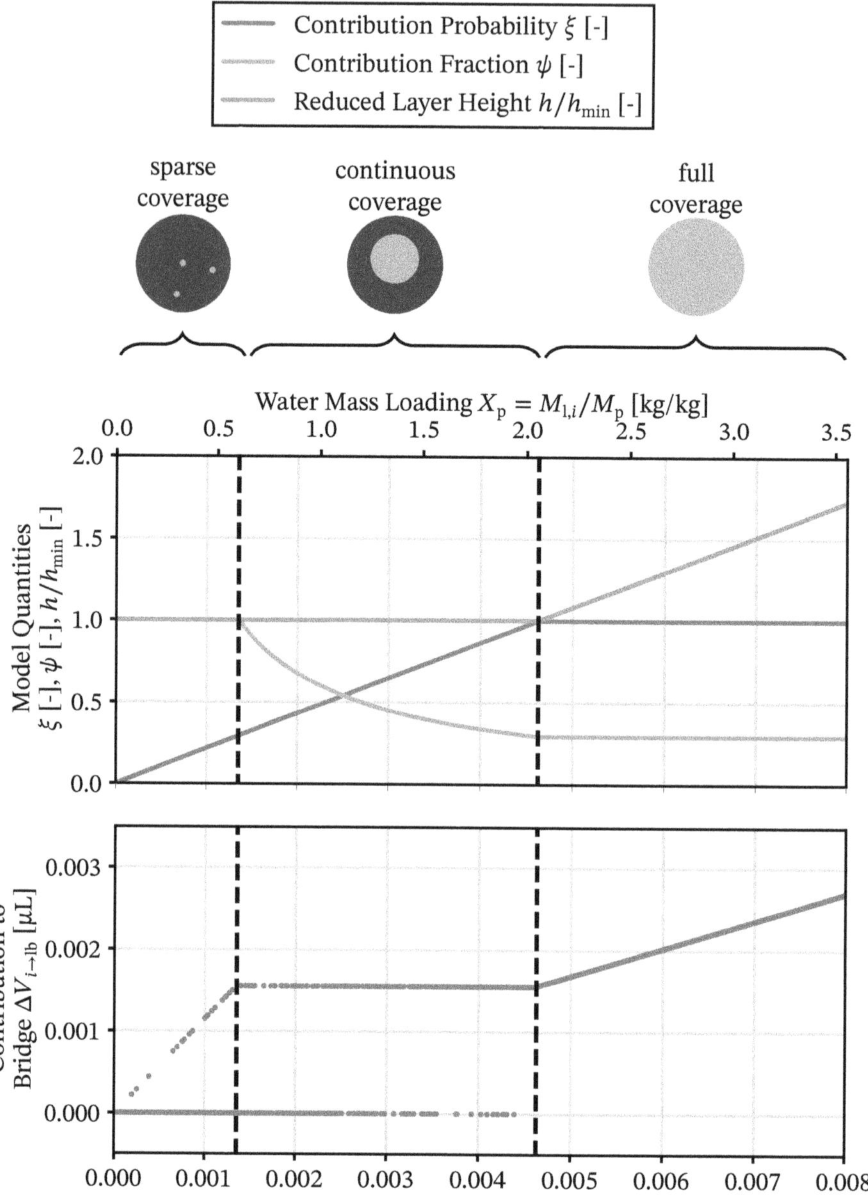

Fig. 3.6.: General characteristics of the coverage model for $r_i = r_j$ and $h_{\min}/r_i = 0.005$. The plot of the normalized contribution to the liquid bridges is depicted with stochastic variations, as expected from the model. The particle diameter here was chosen to be 1.3 mm. The lower graph is dotted due to the stochastic nature of liquid bridge contributions.

3.1.5. Particle Softening Correction and Coarse Graining Approaches

The use of coarse-graining and particle softening has been sparsely treated in literature, despite these methods being integral to the applicability of DEM for wetted systems on the industrial scale. Particle softening allows the use of larger time steps, but increases contact times. Coarse graining techniques reduce the number of particles - and therefore the number of contacts in the system. This requires further adjustments of contact models to retain validity.

Nasato (2016) calibrated a DEM model of a wetted granular system comprising sand and water using a static angle of repose and incline plane setup. He proposed scaling the capillary force law with the square of the scaling factor δ_{CG}^2 and validated this model using a die filling test. This scaling criterion satisfies the conservation of stresses. Analysis for further interaction parameters, like separation distances and bridge rupture, are not given.

Chan and Washino (2018) proposed conserving separation and rupture distances between coarse-grained and original system. They investigated the force scaling of capillary and viscous liquid bridge forces based on a high shear mixer. They found that scaling both of these forces with δ_{CG}^2 preserved velocity distributions as well as mixing conditions in this system. The influence of particle softening on velocity distributions was studied. The velocity distribution was found to be unchanging beyond a particle modulus of elasticity of 100 MPa.

Girardi et al. (2016) and Tausendschön et al. (2020) studied wetted fluidized beds using the CFD-DEM method and applied a filtered drag model. While Girardi et al. (2016) emphasized the importance of filtered drag models for the use of coarse grids, Tausendschön et al. (2020) determined the optimal smoothing length for particle-coarsened simulations when keeping a fine grid. In the course of the study, Tausendschön et al. (2020) confirmed that the choice of l^2 as the scaling factor is applicable in fluidized systems.

Boyce et al. (2017) applied the concept of correcting simulations with softened particles to the liquid bridge forces for the contact phase by applying a factor $\left(Y^*_{\mathrm{sim}}/Y^*_{\mathrm{orig}}\right)^n$, where $n = 0.4$ in the case of a Hertz potential. This reduces the magnitude of the forces to account for the longer contact duration that a lower particle stiffness induces. The consequence of this, however, is a general failure to reproduce forces when they are in a static assembly.

3.2. Liquid Bridge Model Composition and Implementation

To model liquid bridging, one needs to combine a set of the aforementioned force models and closures (see Fig. 3.7), namely:

- capillary force model, providing $\mathbf{F}_{\text{cap}}(V_{\text{lb}}, \delta_{\text{lb}})$,
- viscous force model, providing $\mathbf{F}_{\text{visc}}(, \delta_{\text{lb}})$,
- liquid bridge state model, providing $V_{\text{lb}}(M_{\text{l,i}}, M_{\text{l,j}}, ...)$,
- liquid bridge rupture model $\delta_{\text{lb,rupture}}(V_{\text{lb}})$,
- liquid bridge redistribution model, providing $\{\Delta M_{\text{l},i}, \Delta M_{\text{l},j}\}(r_i, r_j, V_{\text{lb}})$.

For the current work, the capillary force model of Soulié et al. (2006) is combined with the viscous model of Nase et al. (2001) to provide force closures as both of them support poly-disperse systems. This combination is also present in LIGGGHTS under the name *washino/capillary/viscous*, although with a fixed liquid bridge volume proportionally depending on the surface liquid volume of both participating particles and without transfer into a bridge.

The previously described liquid bridge model was implemented in a manner which pre-calculates the layer height as soon as two particles approach each other. An estimate for contributing liquids is then made. The contributed liquid mass of each of the particles is cached for later evaluation, as the stochastical nature of the process would otherwise carry a bias. When the layers touch, the layer height and liquid contributions of both particles is recalculated, and then transferred into a contact-specific liquid bridge volume variable. This variable is tracked and used for subsequent calculations of liquid bridge forces. After the liquid bridge is established, every subsequent step with a liquid bridge volume carries a check for rupture according to the rupture criterion. In this case, the one by Lian et al. (1993) was chosen due to its prevalence in many other studies.

The resulting contact dynamics are shown in Fig. 3.8 by example of a binary contact. The steady line indicates the results of the liquid bridge model and the dotted line of the same situation without using the model. As the particles approach each other in **phase I.**, they follow their inertia or other forces acting upon them. Starting with the liquid layer contact in **phase II.**, the capillary force accelerates the particles towards each other. This is negligibly counteracted by the viscous force.

liquid bridge state

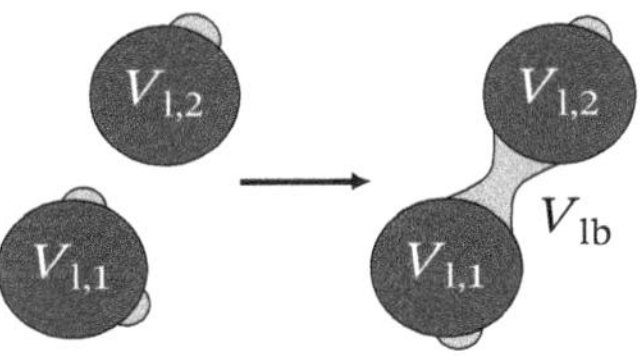

- contribution of each particle to liquid bridge
- determines liquid bond network topology in bulk
- minimum layer height h_{min}

capillary force

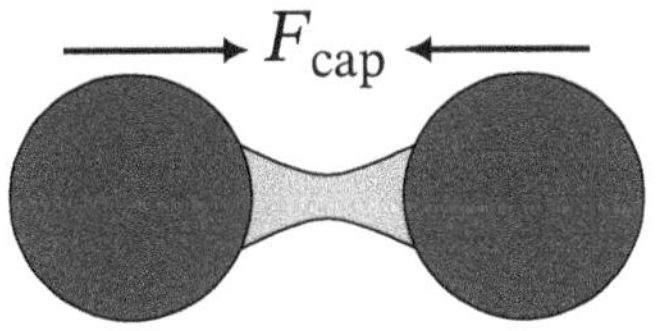

- symmetric potential; acts in normal direction
- only attractive; dissipative action due to bridge stretching
- surface tension $\sigma_{surface}$
- contact angle θ_{eff}

viscous force

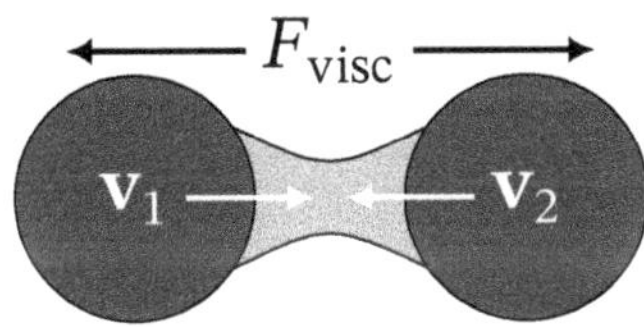

- determines viscous dissipation on impacts
- velocity-dependent; asymmetric potential
- viscosity μ_l
- minimum separation $\delta_{lb,min}$

liquid bridgc rupture

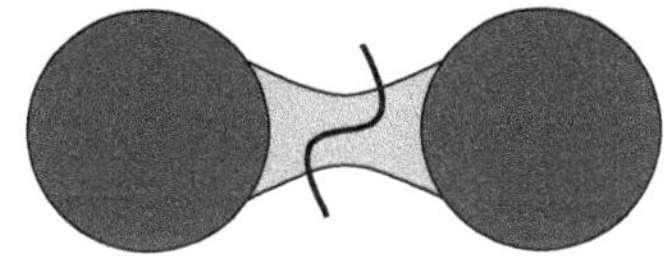

- force cut-off primarily dependent on liquid bridge volume
- determines cohesivity in static loads and dissipation
- contact angle θ_{eff}

Fig. 3.7.: Overview of the four phenomena to be considered in a liquid bridge model and thc individual sensitivities to model parameters and material properties.

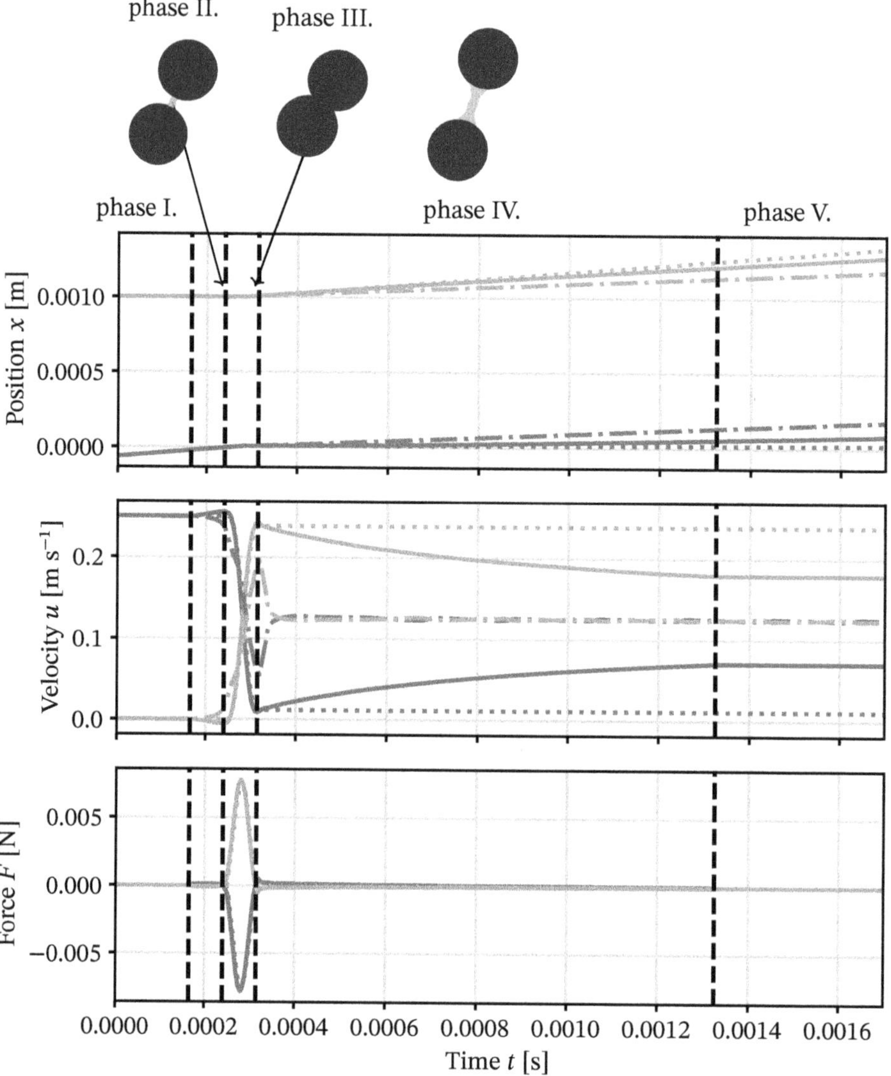

Fig. 3.8.: Position, velocity and force over time for two particles during a binary contact with the implemented liquid bridge model (straight lines) and without (dotted lines). Two glass spheres ($d_p = 1$ mm) were brought into contact at an initial velocity of $u_{p,r,0} = 0.25\,\mathrm{m\,s^{-1}}$ and a volumetric water loading of $0.0034\,\mathrm{m^3\,m^{-3}}$ for both particles.

When the particles come into direct contact in **phase III.**, the contact forces dominate. When comparing the velocities of with and without modeling of liquid bridges during **phase II.** and **III.**, little difference can be seen in the velocities and the forces exchanged. During particle-particle contact (**phase III.**), velocities are low. The effect of viscous forces is thus insignificant with the used minimal separation distance of $\delta_{\mathrm{lb,min}} = 1 \cdot 10^{-6}$ m and low viscosity of the liquid in question (water). As a conservative potential, the capillary force does not play a substantial part during the collision phase itself.

Once the particles are not in contact anymore, the liquid bridge stretches (**phase IV.**). Here, capillary forces play a more important role than the viscous force due to the reciprocal dependence on particle surface distance (compare Eq. (3.16)). The capillary force, in its approximation given by Mikami et al. (1998), falls rapidly with the distance ($F_{\mathrm{cap}} \sim \exp(-\delta_{\mathrm{lb}})$, compare Eq. (3.4)). The negative factor in the exponent is due to the A coefficient in Eq. (3.4) being inherently negative.). The law of Rabinovich et al. (2005) was used in the simulation, but the simpler expression for the relationship between separation distance and capillary force by Mikami et al. (1998) is more suitable for this particular discussion. In **phase V.**, the bridge is broken and liquid-induced cease to act on the particles.

3.3. Experimental Characterization

To characterize a bulk solid, its behavior under a variety of different regimes of flow and consolidation must be assessed. In particular, the sensitivity of the material system has to be extracted and a set of measurement conditions suitable for calibrating is defined. This includes identifying asymptotic behavior and smoothness with respect to the measurement conditions to be able to reliably calibrate a numerical model.

3.3.1. Basic Material Properties

The particle size distribution (PSD) is often too wide to be used for simulation, presenting with an abundance of fines that preclude numerical treatment due to the sheer number of particles. Thus, the fines' tail present in particle size distributions is frequently truncated. This introduces a degree of freedom to the system, making it crucial to calibrate the particle density. This is important because the resulting bulk density would otherwise give a different level of fill in apparatuses if one inserts an identical mass in simulation and experiment. Practically, this test is conducted by filling a vessel of known volume with particles and determining the change in mass.

3.3.2. Shear Testing

A shear test serves the purpose of determining the frictional reaction of pre-consolidated granular matter to shear stresses under pressure. As such, the shear test in its simplest form, shown in Fig. 3.9, consists of granular matter between two plates, one of which is stationary and the other one able to exert a fixed pressure and able to move.

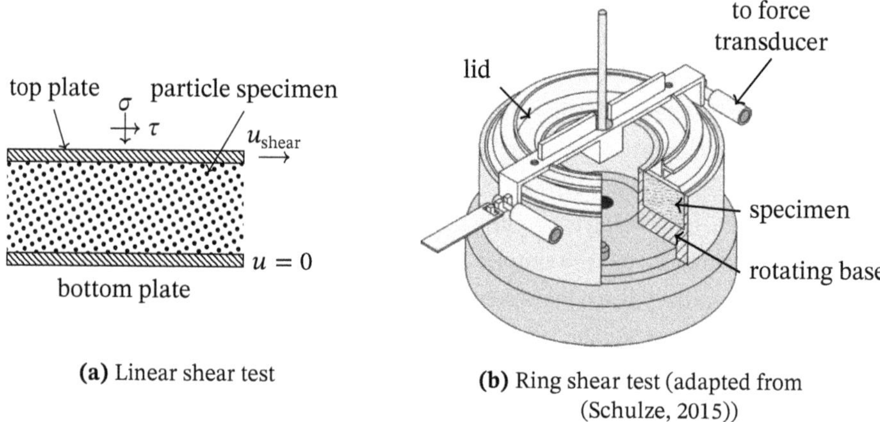

(a) Linear shear test

(b) Ring shear test (adapted from (Schulze, 2015))

Fig. 3.9.: Principle of two apparatuses for shear testing. σ is the normal stress and τ is the shear stress apped. u_{shear} is the shear velocity imposed.

The test is conducted in 3 stages, as illustrated in Fig. 3.10:

1. *pre-consolidation*: A fixed normal stress σ_{preshear} is applied to the upper plate and is moved along the particle rate at a low but constant velocity $u_{\text{plate}} = u_{\text{shear}}$. The particles will begin to flow at a critical point. At this point, the required shear stress will drop rapidly and settle at a lower equilibrium. This state, stationary flow, is characteristic for the normal stress.

2. *relaxation*: The motion of the plate is stopped ($u_{\text{plate}} = 0$) and the pressure reduced to a lower state.

3. *shear to failure*: The plate is set in motion ($u_{\text{plate}} = u_{\text{shear}}$) and the shear stress applied is recorded. The maximum shear stress is the yield point as its subsequent decrease indicates loosening of the material and incipient flow.

For a given normal stress σ_{preshear}, this test is repeated for a set $\{\sigma_{\text{shear},i} < \sigma_{\text{preshear}}\}$ of normal stresses. The output parameter gives a characteristic set $\{(\sigma_{\text{shear},i}, \tau_{\text{shear},i})\}$ that is referred to as the yield locus. For many powders, this set of points can be approximated with a linear relationship $\tau = m\sigma + \tau_0$, where $\Theta_0 = \tan^{-1}(m)$ is also known as the internal angle of friction and τ_0 is the cohesion. Further details on the macroscopic mechanics of bulk solids and the physical background of the yield locus can be found in the reference work of Schulze (2008). Experiments were conducted using a *Schulze RST-XS* ring shear tester (Schulze, 2015) that allows for rapid measurement of any yield locus with little manual interaction.

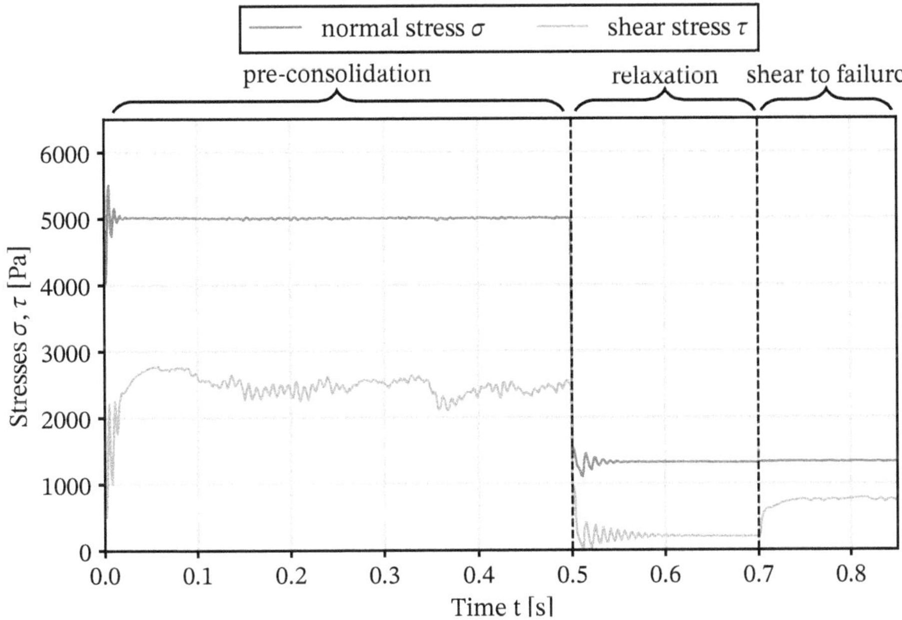

Fig. 3.10.: Normal and shear stresses during shear testing of a granular system.

To this end, a DEM model was developed where a bulk solid specimen is confined in a hexahedral domain between two parallel plates, equipped with baffles to prevent the particles from slipping. The remaining side walls are defined to be treated as periodic boundaries. The normal stress is enforced by the top plate using a *P* controller, while the bottom plate is moved at a constant velocity, as described by Aigner et al. (2013). A purely elastic contact potential is important for this kind of stressing, equating to a coefficient of restitution of $e_{pp} = 1$ and $e_{pw} = 1$. This does not have any further consequences for the test as no kinetic impacts are involved.

3.3.3. Static Angle of Repose

The static angle of repose gives rise to the frictional properties due to irregularities of particle shape and surface roughness. In its simplest form, the angle of repose describes the angle to the horizontal a heap of a granular solid will naturally assume. This is due to the equilibrium between friction and gravitation. There are three options for conducting this test:

1. Discharging a bulk onto a pile through a narrow orifice, and determining the angle of the pile.
2. Discharging a bulk through an orifice, and determining the angle of the remaining bulk.
3. Removing vertical boundaries, shown in Fig. 3.11.

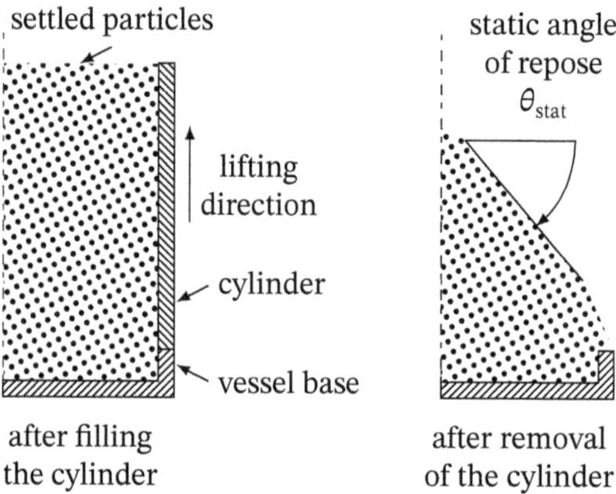

Fig. 3.11.: Principle of an exemplary static angle of repose measurement. θ_{stat} is the resulting static angle of repose.

For powders, the former method is documented in DIN ISO 4324. In practice, this method is limited in its responsiveness to very cohesive systems. Ensuring a flow of cohesive, possibly coarse particles through a narrow orifice often poses an issue, limiting the usability of the test. Therefore, the second option was chosen and implemented.

3.3.4. Dynamic Angle of Repose

While a shear test and the static angle of repose should be able to predict the dynamic angle of repose, it is beneficial to have a test that covers shear behavior over a wide range of stresses. This predestines the dynamic angle of repose that occurs in a drum for this purpose as a wide range of stresses occur inside the apparatus (Fig. 3.12) as part of the flow of the bulk. Regimes changes can be easily induced by changing the fill level and rate of rotation. In particular, viscous forces can be observed as kinetic impacts happen at higher rotation rates. This makes this test sensitive not only to capillary forces (comprising the influence of surface tension, contact angle and a model parameter for the bridging state), but also viscous forces that increase with velocity and decrease with a model parameter for the asperities on the surface.

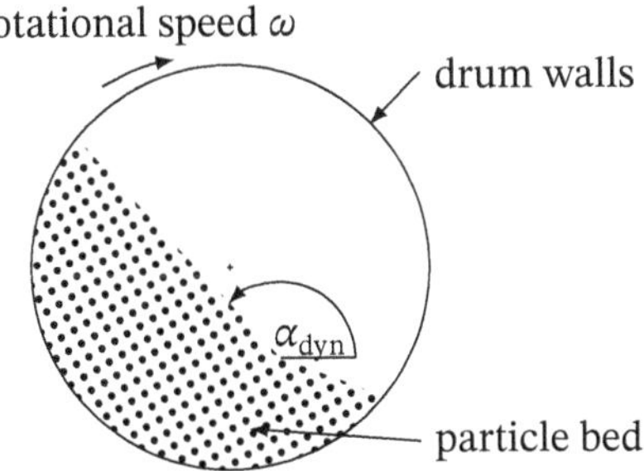

Fig. 3.12.: Principle of a dynamic angle of repose in a rotating drum. ω indicates the angular velocity of the drum and Θ_{dyn} the resulting dynamic angle of repose.

To determine an actual angle from simulation data that is present in the form of $\{x_i, y_i, z_i\}(t)$ for each particle i at every time t, the most accurate way is the mapping of this data to a two-dimensional histogram $q_0(x, y)$ with respect to the vertical direction y and horizontal direction x. This removes dependence on both time and the actual number of particles in the system. To remove noise, the histogram is then thresholded to create a binarized two-dimensional distribution:

$$q_b(x, y) = \begin{cases} 1 & q_0(x, y) > N_{thr}, \\ 0 & \text{otherwise.} \end{cases} \tag{3.23}$$

The noise in this case comprises disruptions caused by single particles as well as irregularities in particle packing density. The Canny (1986) edge detection algorithm is applied to this and returns the set of points found to be an edge. By choosing a fixed window on the system which is to be observed, one may make the assumption of monotonicity of the outer outline of the particle bulk. Sorting the edge points by their horizontal components and asserting these assumptions will remove any interior details such as the inner vacuole created by the cascading bulk. To find the angle of repose θ, we thus formulate an objective function

$$\begin{aligned} J_{\text{dynamicAOR}}(\{x_i^{\text{edge}}, y_i^{\text{edge}}\}, m, b, i_{\min}, i_{\max}) = & \sum_{i=i_{\min}}^{i_{\max}} (m x_i^{\text{edge}} + b - y_i^{\text{edge}}) \\ & + \lambda_m (\arctan(m))^{-1} \\ & + \lambda_i \left((i_{\max} - i_{\min})^2 (1 + m^2)\right)^{-1} \end{aligned} \tag{3.24}$$

that considers three contributions:

1. *regression & windowing*: $\sum_{i=i_{min}}^{i_{max}} (mx_i^{edge} + b - y_i^{edge})$ ensures that the detected edge points approximate the line equation $y_i = mx_i + b$ between $x_{i_{min}} \leq x_i \leq x_{i_{max}}$, effectively introducing a windowing function. The windowing is required because not every point the edge detector identifies will contribute to the angle of repose.

2. *steepest slope*: $(\arctan(m))^{-1}$ is introduced to select the region with the steepest slope. To balance this contribution, the coefficient λ_m is introduced. Larger values of m minimize the value of this term.

3. *line length*: $\left((i_{max} - i_{min})^2 (1 + m^2)\right)^{-1}$ penalizes overly small data windows. It also considers the slope because the function fit will naturally contain less points $N_i = i_{max} - i_{min}$ for steeper slopes.

Minimizing this function is not a trivial matter due to the presence of many local minima in many dimensions. Furthermore, it is a constrained optimization because a minimum window size $N_{i,min} >= (i_{max} - i_{min})$, among other windowing constraints ($i_{min} \geq 0$, $i_{max} \leq N_{i,max}$), has to be fulfilled. Thus, a basin-hopping optimization algorithm according to Wales and Doye (1997) that is implemented in the SciPy library (Virtanen et al., 2020) was used to find a global minimum of this objective function. The parameters had to be extensively tweaked to give values that represent the system at hand.

3.4. Sensitivity Analysis of the Numerical Model

There are a wide range of material and model parameters that must be adjusted and calibrated to ensure physically consistent predictions. In this section, the influence of non-intuitive model parameters on typical calibration experiments will be analyzed to find bounds for these. In particular, the minimum layer height h_{min} will be analyzed because of the unknown performance of the bridge state model.

3.4.1. Restitution Analysis

Particle-particle contacts can be directly characterized using restitution analysis, where two particles are brought into contact with an initial relative velocity. The key output parameter is the resulting coefficient of restitution (COR) $e = u_{r,final}/u_{r,initial}$ that commonly has a value $e < 1$ due to dissipation. This dissipative character of particle collision has been studied for wetted particle systems to a great extent by, for example, Buck et al. (2017) and Antonyuk et al. (2009).

The influence of the minimum layer height model on the coefficient of restitution is shown in Fig. 3.13, where the collision of two wetted particles is evaluated. The higher the minimum layer height parameter is chosen, the stronger the liquid bridge forces become. This becomes very apparent at liquid concentrations larger than $V_{l,i}/V_p = 0.0112$, as the sticking point, where the coefficient of restitution is zero, deviates by more than $0.1\,\mathrm{m\,s^{-1}}$ when looking at the range of reasonable values for the minimum layer height. Not considered in this study is the influence of the reduced probability of contacts occurring for a given liquid loading.

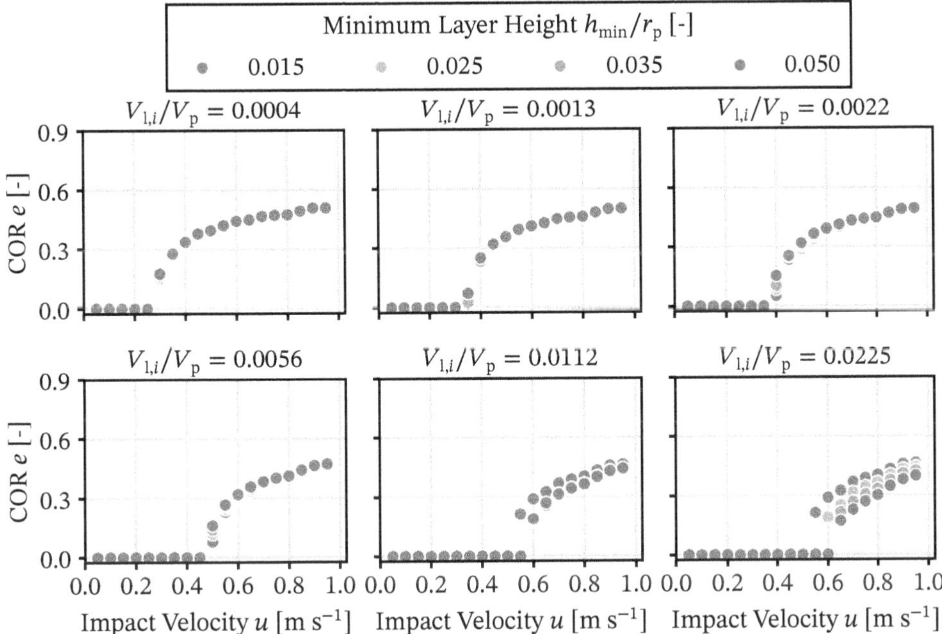

Fig. 3.13.: Model sensitivity with respect to minimum liquid layer height and particle surface liquid volume, given a normal impact of two wetted particles with a density of $\rho_p = 2250\,\mathrm{kg\,m^{-3}}$, a minimum separation distance of $\delta_{lb,min} = 1 \cdot 10^{-6}\,\mathrm{m}$, a modulus of elasticity of $Y_p = 5 \cdot 10^7\,\mathrm{Pa}$ and a dry coefficient of restitution of $e = 0.9$.

The coefficient of restitution resulting from wet particle-particle collision shows a strong sensitivity to the minimum separation distance used in the viscous force model, as shown in Fig. 3.14. Kinetic impacts dissipate more energy when the minimum separation distance is smaller. For example, the characteristic of the restitution test, the sticking point u_{stick} increases from $u_{stick} < 0.5\,\mathrm{m\,s^{-1}}$ for a minimum separation distance ratio $\delta_{lb,min}/r_p$ of 0.002 and 0.005 to a value of $u_{stick} = 0.6\,\mathrm{m\,s^{-1}}$ for a minimum separation distance ratio of $\Delta_{lb,min}/r_p = 0.01$ for a volumetric liquid loading $V_l/V_p = 0.0225$.

Generally, the trend is that:

- higher liquid loadings V_l/V_p decrease the sticking velocity u_{stick},
- lower minimum separation distances $\delta_{lb,min}$ (ratios) increase the sticking velocity u_{stick}.

This is due to $\delta_{lb,min}$ limiting the contribution of the viscous force during impact - the law of Nase et al. (2001) gives a relationship of $F_{visc} \sim \log(r_p/\delta_{lb,min})$ during collisions. This fits the experience of Antonyuk et al. (2009), who found this parameter to have a large influence in modelling particle-wall impacts.

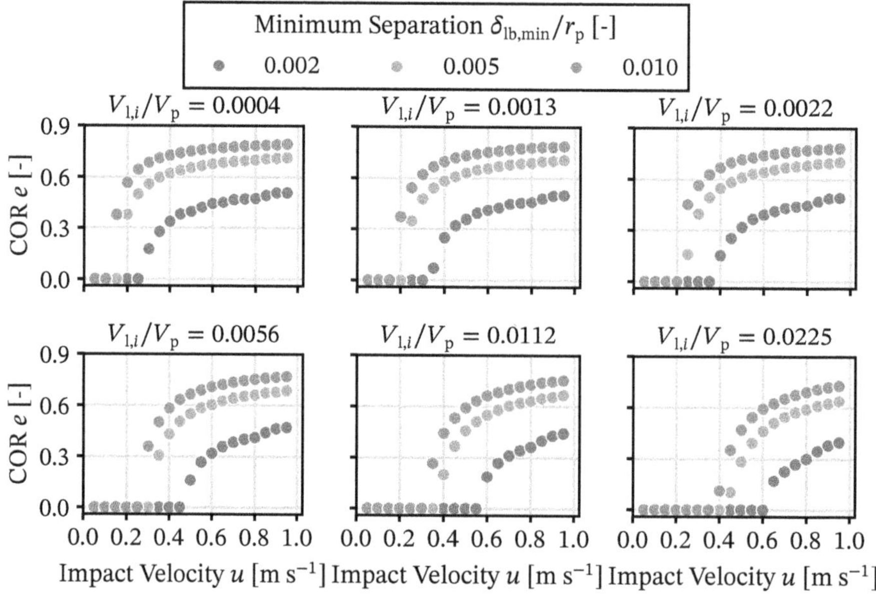

Fig. 3.14.: Model sensitivity with respect to minimum separation distance and particle surface liquid volume, given a normal impact of two wetted particles with a density of $\rho_p = 2250\,\mathrm{kg\,m^{-3}}$, a minimal liquid layer height of $\Delta h_{min} = 1 \cdot 10^{-6}\,\mathrm{m}$, a modulus of elasticity of $Y_p = 5 \cdot 10^7\,\mathrm{Pa}$ and a dry coefficient of restitution of $e = 0.9$.

3.4.2. Sensitivity of the Static Angle of Repose

As kinetic effects do not play a substantial role in statically loaded, wetted scenarios, restitution experiments cannot capture the effects at work.

One example that captures this scenario is the bulk angle of repose test that will be discussed in section 3.3.3. Due to the absence of relative velocities, the viscous force $F_{\text{visc}} \sim v_{\text{n}}$ does not play a role here. The key parameter is the angle that develops as a consequence of the capillary force F_{cap}. As an example, the leftmost column of Fig. 3.16 shows the angle of repose that develops depending on the water loading $X_{\text{p}} = M_{\text{water}}/M_{\text{p}} = \rho_{\text{water}}V_{\text{water}}/\rho_{\text{p}}V_{\text{p}}$ in the bulk. The heaps that form in the experiments show a much higher increase in angle of repose going from $X_{\text{p}} = 0.2\,\text{g}\,\text{kg}$ to $X_{\text{p}} = 0.4\,\text{g}\,\text{kg}^{-1}$ than going from $X_{\text{p}} = 0.4\,\text{g}\,\text{kg}^{-1}$ to $X_{\text{p}} = 0.6\,\text{g}\,\text{kg}^{-1}$. Applying the cohesion law proposed in 3.2 with a very low value of $h_{\text{min}} = 2 \cdot 10^{-7}\,\text{m}$ equates to applying the *washino/capillary/viscous* model with a surface liquid-contribution of $\psi_{\text{max},i} \approx 13\%$, as given by Eq. (3.20). The results of this study, depicted in Fig. 3.16, show very little responsiveness of the system to the liquid loading. This is unsurprising, since capillary force laws agree that the capillary force is independent of the amount of liquid present in the bridge (see Fig. 3.2). To capture this important effect, the heterogeneity of liquid distribution within contact networks has to be taken into account - a distinguishing feature of partially-wetted solids. The analyzed static angles of repose are given in Fig. 3.15 and Fig. 3.16 depicts snapshots of the experiments and simulations of the test.

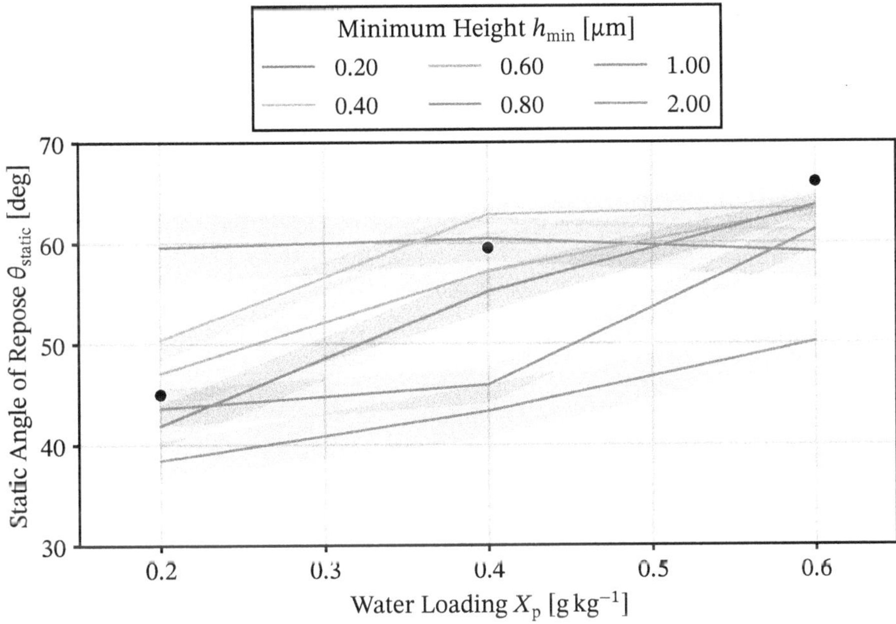

Fig. 3.15.: Influence of the minimum layer height liquid bridge model parameter h_{min} on the static angle of repose θ_{static}. The black circles indicate experimental data points.

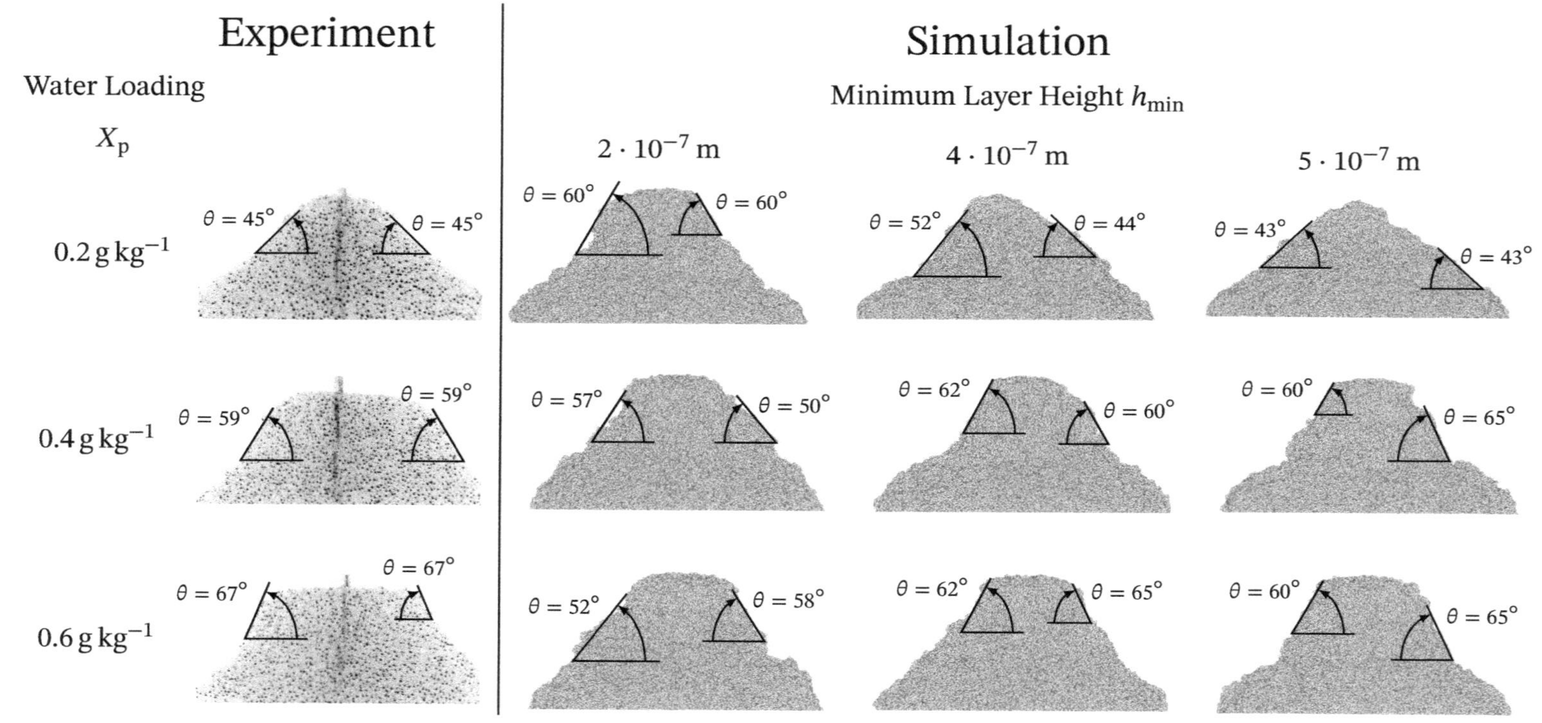

Fig. 3.16.: Example angle of repose test for demonstrating the effect of modulating capillary bridge networks using the liquid bridge state model. Water was used as a liquid and the particles are glass particles with a mean diameter of $d_p = 1.3$ mm. Note that the limited height of the cones is due to the limited amount of material that was used for the experiments and is not indicative of cohesive behavior. Experimental images were captured with a camera and the colors were inverted.

As expected, a higher minimum layer height h_{min} reduces the overall angle of repose by reducing the number of liquid bridges in the system. On the other hand, adding more water (increasing the liquid loading X_p) counteracts this effect in the model. This simultaneously reduces the volume that each liquid bridge can carry. For low water loadings ($X_p = 0.2\,\mathrm{g\,kg^{-1}}$), higher minimum layer heights, exemplified by $h_{min} = 5 \cdot 10^{-7}$ m, show a weak increase in the static angle of repose over the dry state. This is due to the low percentage of liquid present. Saturation can be seen for $h_{min} = 2 \cdot 10^{-7}$ m where the angle of repose does not change significantly from 60° when adding more water.

While the minimum layer height h_{min} has a modulating influence on the topology of contacts that are deemed wetted, the surface tension $\theta_{surface}$ has a direct, linear influence on the capillary force acting in these contacts $F_{cap} \sim \theta_{surface}$. As the static angle of repose test is dominated by the capillary force, this influence must be analyzed, as is done in Fig. 3.17. While the surface tension of the actual liquid $\sigma_{surface} = 0.0728\,\mathrm{N\,m^{-1}}$ slightly underpredicts the angle of repose, increasing it by about 35 % to $\theta_{surface} = 0.1$ Pa s reproduces the angle perfectly for $h_{min} = 1 \cdot 10^{-6}$ µm. This is especially apparent at a water loading of $X_p = 0.4\,\mathrm{g\,kg^{-1}}$. Here, the probability of creating liquid bridges is identical, whereas the increase in liquid bridge strength leads to a large increase in the angle of repose.

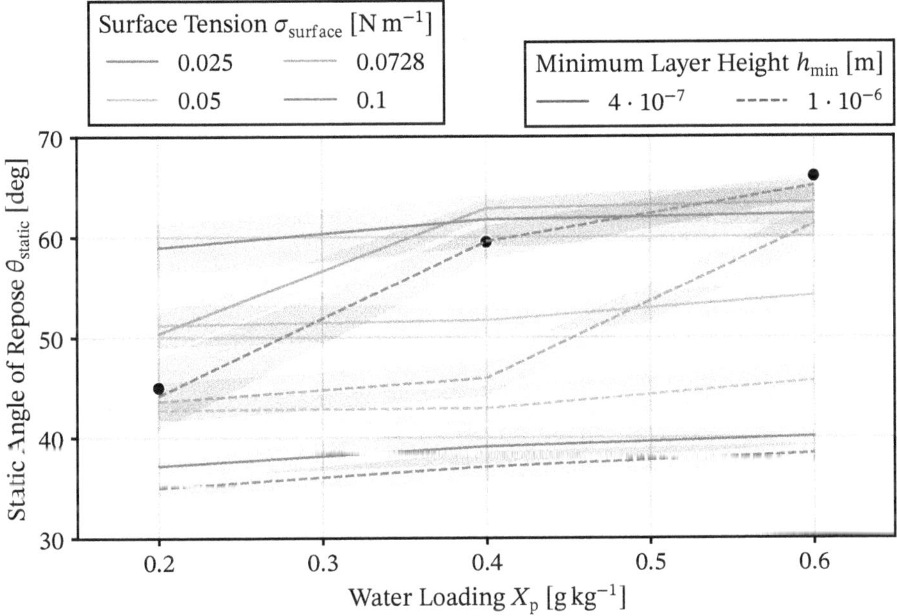

Fig. 3.17.: Influence of the surface tension on the static angle of repose θ_{static}. The black circles indicate the mean angle of repose.

3.4.3. Sensitivity of the Shear Test

The shear test, as well as the static angle of repose test, is sensitive to the static properties of liquid bridges rather than their dynamic behavior.

The experimental data for the friction angle and the Coloumb cohesion shows little systematic response in the regime surveyed in the experimental data. None of the chosen parameters showed a satisfactory agreement between simulation and experiment - most likely owed to the low reproducibility of the experiment. As such, the focus is laid on reproducing the trends present. Varying the minimum layer height parameter (Fig. 3.18) does not lead to any obvious change in the friction angle with increasing water content, in accordance with the experimental data. The cohesion τ_c, however, generally increases for $0.4\,\mathrm{mm} \leq h_{\mathrm{min}} \leq 1\,\mathrm{mm}$, the range of values that provide a compromise between bridge topology and bridge volume. This trend reflects the experimental behavior.

The surface tension $\theta_{\mathrm{surface}}$ (Fig. 3.19) does not change the friction angle either. As expected, as the capillary force, obeying $F_{\mathrm{cap}} \sim \theta_{\mathrm{surface}}$, only acts in the normal contact direction. This ignores rolling friction. For assemblies with low coordination numbers and attractive normal forces, as well as rolling friction, the accuracy of this statement has to be reconsidered due to "carry-over" into the tangential direction. The Coulomb cohesion τ_c increases with higher surface tension, in accordance to the previous argument.

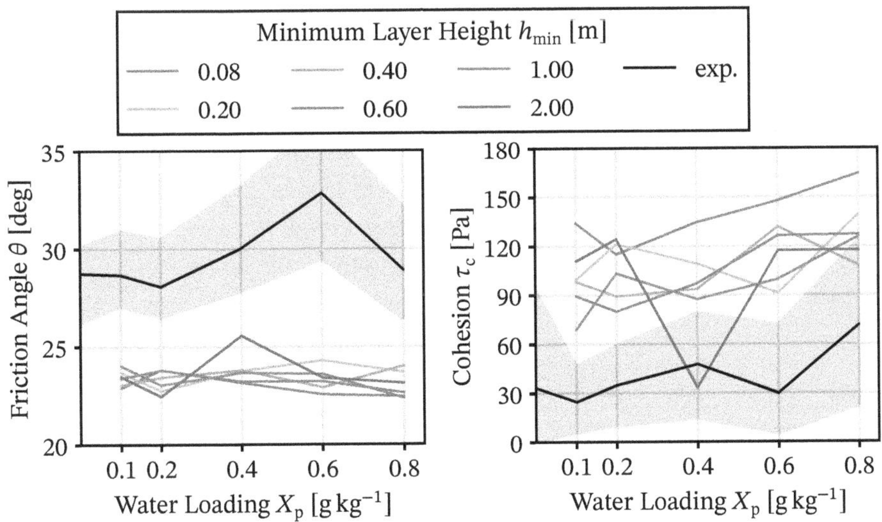

Fig. 3.18.: Influence of the minimum layer height parameter h_{min} on the friction angle θ and the Coloumb cohesion τ_c. The experimental results are given in black. The shaded regions indicate the standard deviation of the data. The consolidation stress was set at $\sigma_{pre} = 5000\,\text{Pa}$ and a surface tension of $\sigma_{surface} = 0.0728\,\text{Pa s}$ was used.

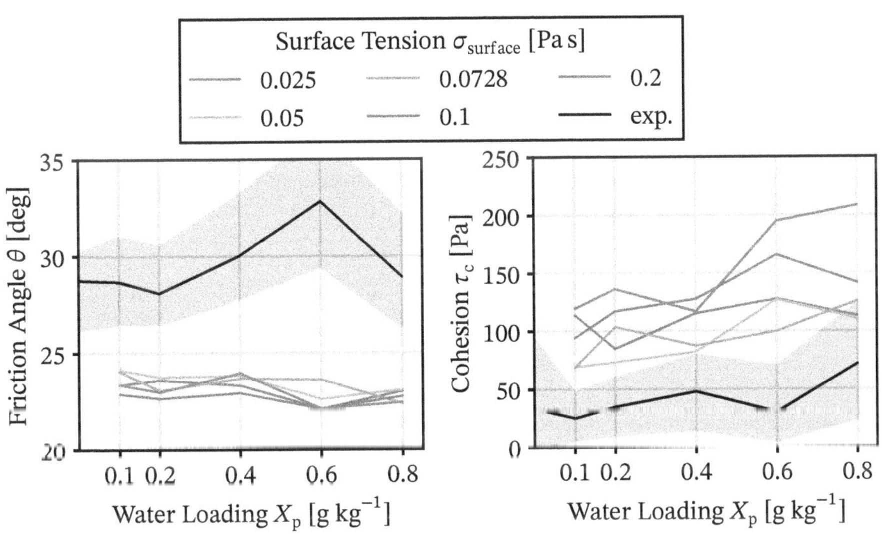

Fig. 3.19.: Influence of the surface tension $\theta_{surface}$ on the friction angle θ and the Coloumb cohesion τ_c. The experimental results are given in black. The shaded regions indicate the standard deviation of the data. The consolidation stress was set at $\sigma_{pre} = 5000\,\text{Pa}$ and a minimum layer height of $h_{min} = 1\,\mu\text{m}$ was used.

3.4.4. Sensitivity of the Dynamic Angle of Repose

As the dynamic angle of repose is a dynamic quantity, the parameters of interest extend to those influenced by the viscous force law, namely the liquid viscosity μ_l, the minimum separation distance $\delta_{lb,min}$, as well as the other parameters. Of course, the minimum layer height h_{min} has an influence on the dynamic angle of repose. In general, it is expected that the angle of repose will decrease with higher values of h_{min}. This trend could be observed in the simulations as shown in 3.20 and was most pronounced for $h_{min} < 1\,\mu m$; $h_{min} \geq 1\,\mu m$, owing to the diminishing probability of liquid bridging.

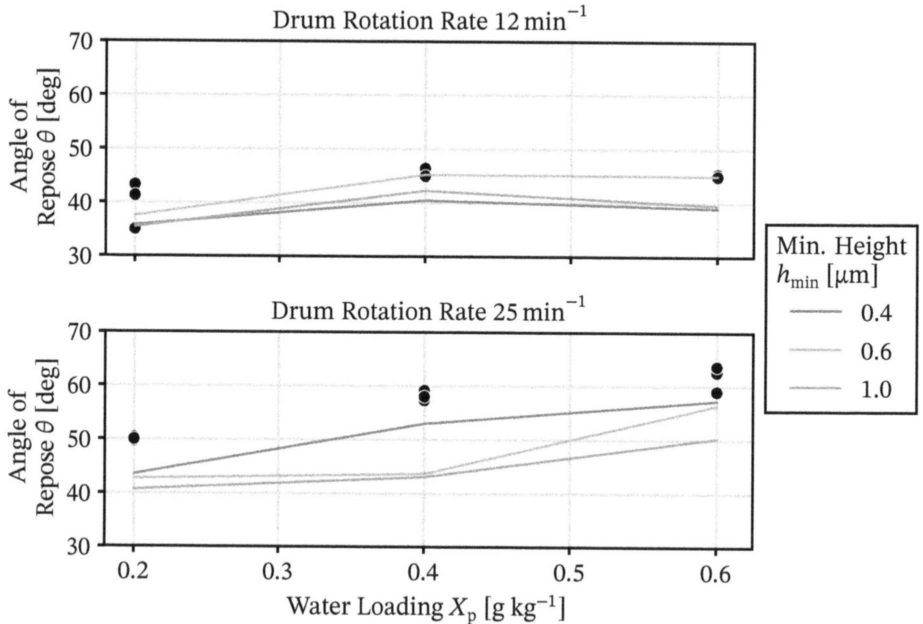

Fig. 3.20.: Influence of the minimum layer height parameter on the dynamic angle of repose. The experimental data is indicated in black. The simulations were run with $\sigma_{surface} = 0.1\,N\,m^{-1}$, $\delta_{lb,min} = 656\,\mu\,m$, $\mu_l = 0.000\,89\,Pa\,s$.

Varying the surface tension $\sigma_{surface}$, shown in Fig. 3.21, results in steeper angles with larger values of $\sigma_{surface}$. In some scenarios, a drop is indicated. This is an artifact of the analysis algorithm that does not handle chunky flows well. Because the surface tension directly modulates the strength of the capillary force F_{cap}, the cohesion of the bulk increases and from some point on, this translates to the transport not of individual particles, but agglomerates. Irregular tesseract motion is difficult to detect using the algorithm applied. For example, at a drum rotation rate of $12\,min^{-1}$ and a surface tension of $\sigma_{surface} = 0.15\,N\,m^{-1}$, the angle of repose exceeds the one of lower surface tension by more than 10° at water loadings $X_p \leq 0.4\,\mathrm{g\,kg^{-1}}$. At $X_p = 0.6\,\mathrm{g\,kg^{-1}}$, the aforementioned agglomeration effect on the analysis algorithm takes over. Higher rotation rates of $25\,\mathrm{min^{-1}}$ overcome this phenomenon due to the higher forces acting in this flow regime.

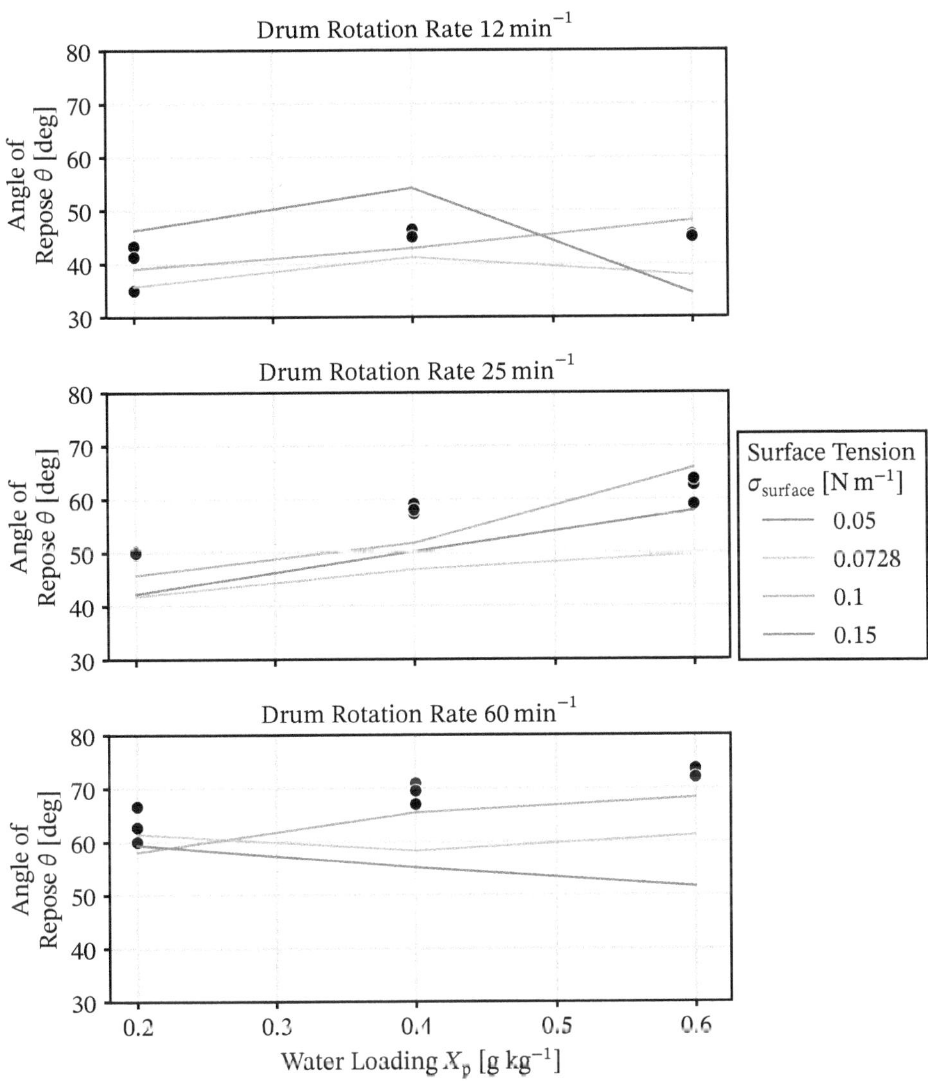

Fig. 3.21.: Influence of the surface tension on the dynamic angle of repose at different rates of rotation. The experimental data is indicated in black. The simulations were run with $h_{min} = 1\,\mu\mathrm{m}$, $\delta_{lb,min} = 656\,\mu\mathrm{m}$, $\mu_l = 0.000\,89\,\mathrm{Pa\,s}$.

When considering the effect of viscosity (Fig. 3.22), a smaller value of the minimum separation distance $\delta_{lb,min}$ has to be chosen, as these parameters correlate. Larger values for the viscosity dampen the velocity of particles during impacts. Instead of directly sticking together, the viscosity decreases the kinetic energy of the particles during impacts to a degree that the capillary force may cause them to agglomerate. As can be expected, the effects of viscosity do not appear at a drum rotation rate of $12\,\mathrm{min}^{-1}$ but at larger values of $25\,\mathrm{min}^{-1}$ where chunking starts to set in at $X_p \geq 0.4\,\mathrm{g\,kg}^{-1}$ and viscosities $\mu_l >= 0.75\,\mathrm{Pa\,s}$ (although it has to be noted that this data is very noisy). For larger rates of rotation, the effect is diminished due to the issues of chunking.

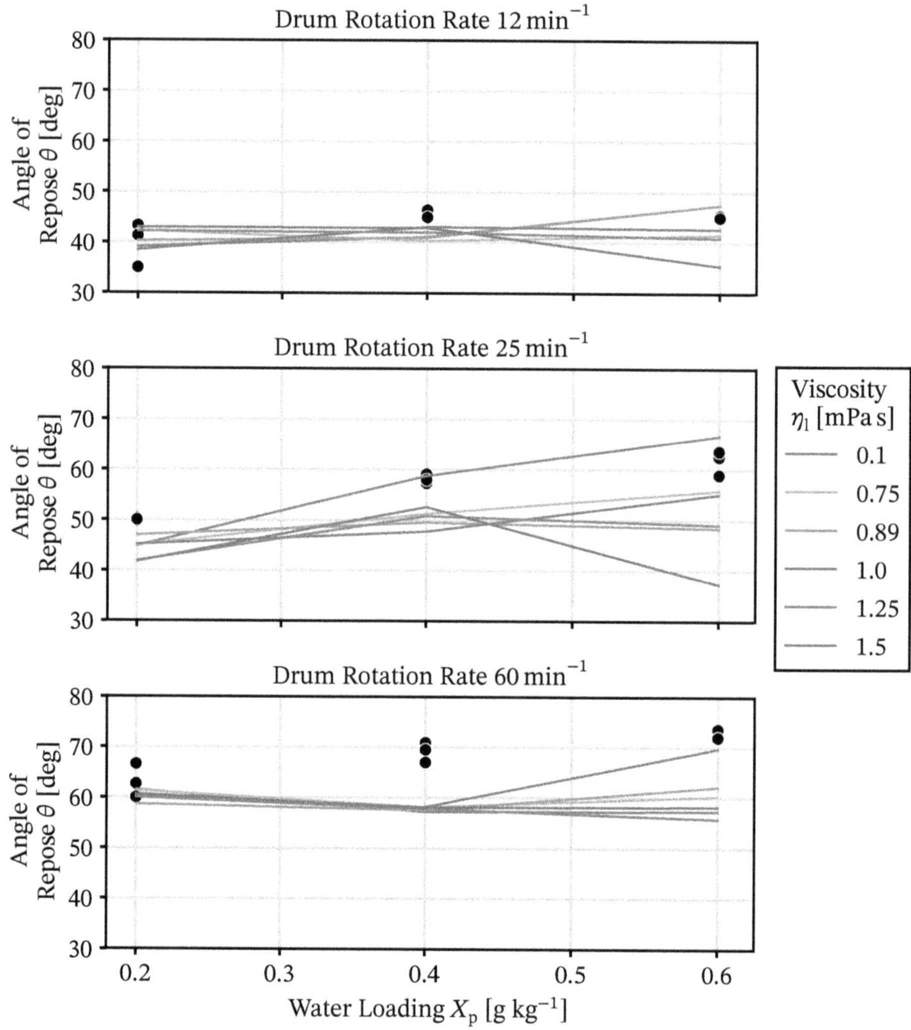

Fig. 3.22.: Influence of the liquid viscosity on the dynamic angle of repose. The experimental data is indicated in black. The simulations were run with $h_{min} = 1\,\mu\mathrm{m}$, $\delta_{lb,min} = 6\,\mu\mathrm{m}$, $\sigma_{surface} = 0.0728\,\mathrm{N\,m}^{-1}$.

Due to the interplay of the viscosity with the minimum separation distance, its influence has to be investigated (Fig. 3.23) by fixing the value for the viscosity. For the lowest rotation rate, only the smallest minimum separation distances ($\delta_{\mathrm{lb,min}} = \{3; 6\}\mu\mathrm{m}$) show a response, indicating that low values of this parameter suppress any influence. At 25°, intermediate values of the range surveyed ($\delta_{\mathrm{lb,min}} = \{32; 65\}\mu\mathrm{m}$) showed the best reproduction. The complex interplay of phenomena highlights the need for comprehensive calibration of all parameters involved.

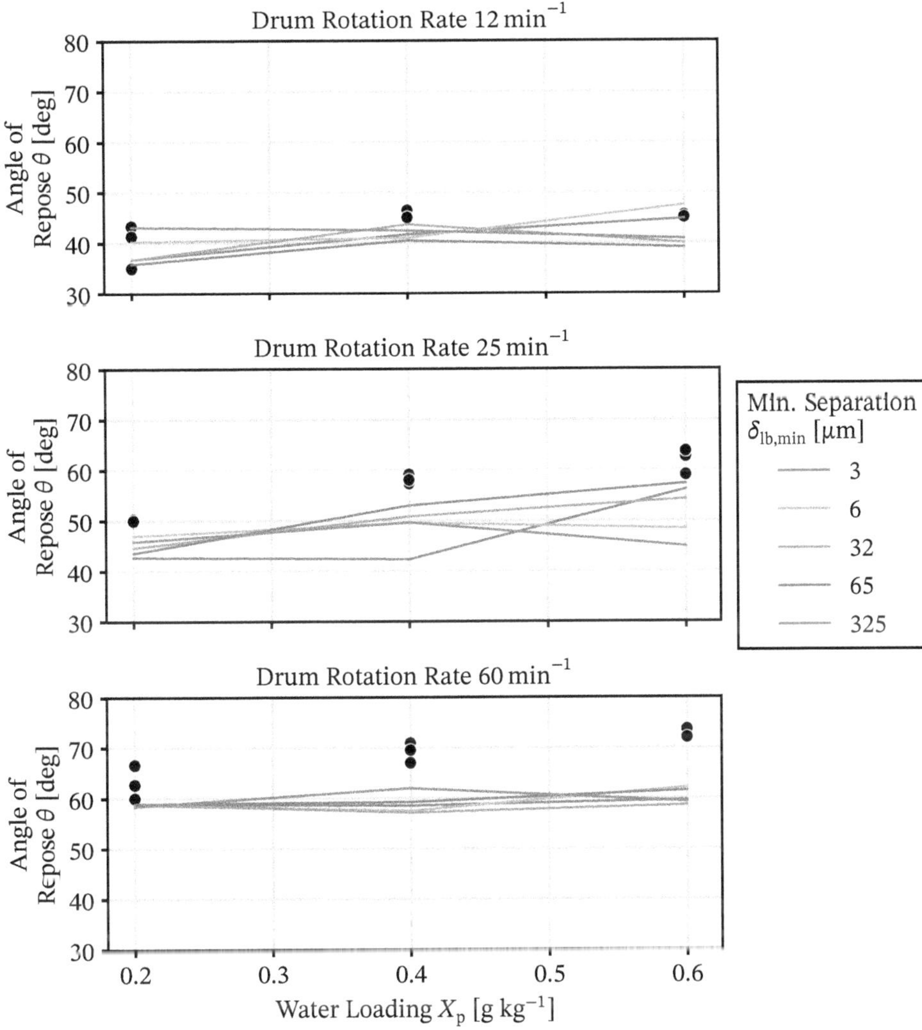

Fig. 3.23.: Influence of the minimum separation distance $\delta_{\mathrm{lb,min}}$ on the dynamic angle of repose. The experimental data is indicated in black. The simulations were run with $h_{\mathrm{min}} = 1\,\mu\mathrm{m}$, $\mu_l = 0.000\,89\,\mathrm{Pa\,s}$, $\sigma_{\mathrm{surface}} = 0.0728\,\mathrm{N\,m^{-1}}$.

3.5. Calibration Workflow for Liquid Bridge Modelling in the Discrete Element Method

The calibration of contact models for use in the discrete element method has long been an area of active research. In the general literature survey of Coetzee (2017), two kinds of approaches are distinguished:

- *bulk calibration*: the reproduction of bulk behavior is seen as the key criterion rather than the function of the contact law within the physical mechanism.
- *direct calibration*: contact model parameters are calibrated according to the physical mechanism involved.

In the works Coetzee (2017) surveyed, bulk calibration was found to be performed with one of the following methods:

- *penetration test*: penetration of a particle bed by a rigid object
- *shear test*: shear stress reaction of a confined element of particles to imposed velocity
- *static angle of repose*: formation of a pile of material under the influence of gravity
- *dynamic angle of repose*: formation of an interface under the influence of gravity in a rotating vessel
- *hopper outflow*: material is allowed to flow out of an orifice under the influence of gravity
- *uni-/bi-/tri-axial compression test*: confinement of a speciment and application of normal stresses

The most popular bulk calibration experiments were the static angle of repose and the shear test. Lessons learned for shear testing are that

- the shear rate can be greatly increased without loss of precision (Härtl and Ooi, 2008),
- increasing the particle stiffness/modulus of elasticity increases inter-particle friction for a given coefficient of friction (Simons et al., 2015).

For the dynamic angle of repose test, it was found that

- both sliding and rolling friction influenced the resulting dynamic angle of repose (Combarros, 2014), and
- the coefficient of restitution and the modulus of elasticity do not affect the dynamic angle of repose (Combarros, 2014).

No literature was found for the calibration of wetted systems in particular. Roy et al. (2016) proposed a form of contact law and a procedure for determining coefficients based on a shear test. While this simplified approach is very tempting, it ignores other features of wetted granular flow like contact network homogeneity.

Due to the broad range of parameters that are available, calibrations have to be performed with smart optimization in mind. The following principles need to be obeyed for an efficient calibration method that this workflow shall obey:

- calibrate model parameters over material properties,
- segregate and subdivide into as many calibration steps as possible, and
- ensure validity for corner cases.

To this end, the workflow that is outlined in Fig. 3.24 is developed. For all other calibration steps to be valid, the bulk density has to be reproduced in simulations. This is particularly important because the particle size distribution might be truncated for simulation efficiency. This step ensures that elementary aspects, such as fill level of the dynamic angle of repose experiment, are given correctly. By setting both the Youngs modulus and particle size distribution first, the computational efficiency of further steps can be ensured, as a low Youngs modulus allows for large time steps to be used and a truncated particle size distribution reduces the number of particles to be considered. The solids/particle density has to be iteratively calculated by guessing an initial value, recreating the container fill in a simulation and then determining the bulk density, and then correcting it.

Alternatively, data from a shear test and a shear test simulation can be used. Calibrating the frictional coefficients for particle-particle interaction has to be performed with simultaneous consideration of static and rolling friction. All three previously described experimental setups are suited to characterize this behavior, with the dynamic angle of repose and shear test being judged to be more resilient towards pre-consolidation compared to the static angle of repose. Simulations for each combination of frictional coefficients have to be performed once and experiments should be repeated to account for possible inconsistencies. The best possible - or rather a sufficiently accurate - parameter set can be determined either using an optimization method or by screening the parameter space at a sufficient resolution and identifying Pareto optimal (Pareto, 1894) parameter combinations. The latter case is only possible when the parameter combination dimensionality is low due to the computational demand imposed by performing the simulations.

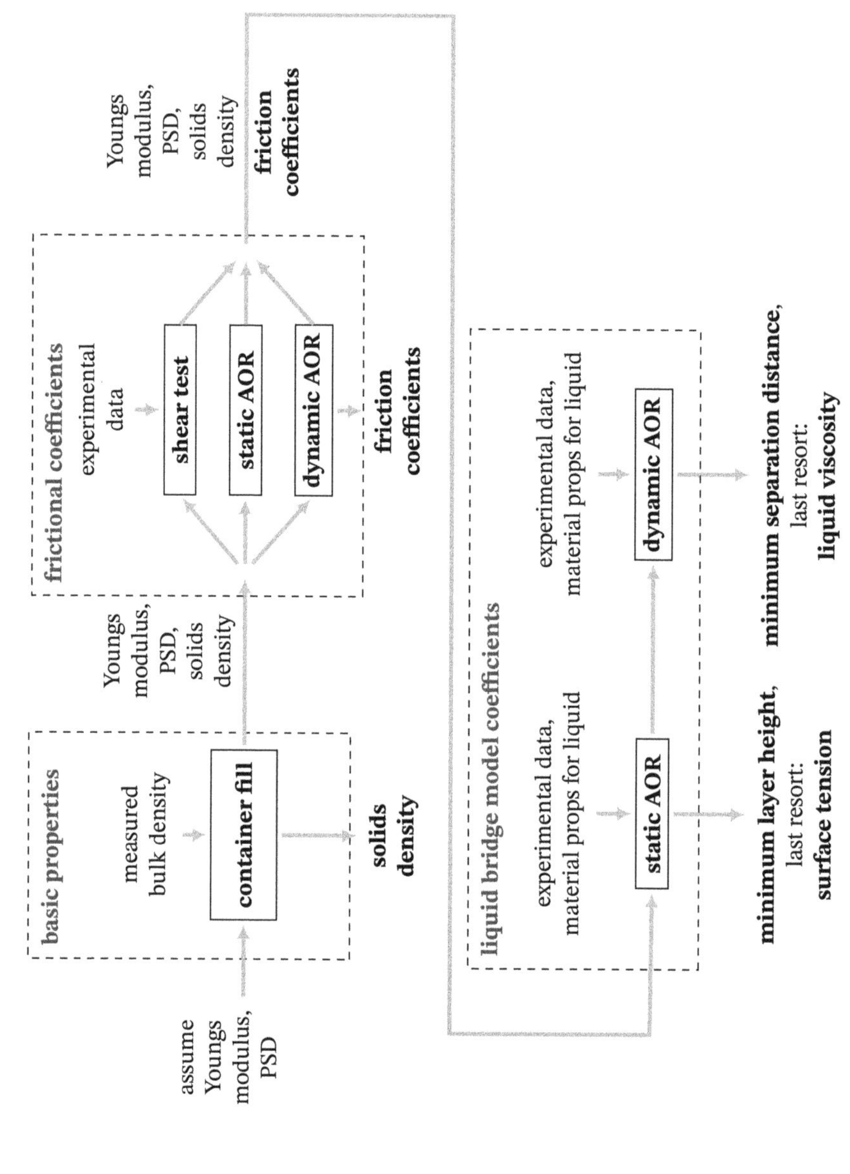

Fig. 3.24.: Schematic depiction of the calibration workflow.

Going onward to the liquid bridge model parameter calibration, only the static and dynamic angle of repose setups were found to be reliably sensitive to the water content. Because the static angle of repose is primarily sensitive to the capillary force and the dynamic angle of repose to both the capillary force and the viscous force, a sequential calibration of first the capillary force and the viscous force second, the calibration can be further subdivided into two stages: first the capillary contribution for the static effects and then the viscous contribution to cover the remaining dynamic effects not explainable by the capillary force.

The primary parameter that will be calibrated in the first stage is the minimum layer height. This modulates the density of the liquid bridge contact network. The secondary parameter is the surface tension which modulates the strength of the liquid bridges that are formed. The minimum layer height is a model parameter and the surface tension is a physical material property. The former should be preferentially used to converge on the desired behavior and the latter as a last resort. In utilizing this workflow, a set of contact model coefficients, which are valid in a wide range of particulate flow regimes, are obtainable.

4 Calibration of a Discrete Element Method Contact Law for Describing a Wetted Granular System

The ultimate goal of the developed calibration approach is the description of arbitrary, real powders in their weakly wetted state. These systems are highly complex due to surface phenomena that take place. Therefore, to demonstrate and develop the workflow, glass particles were chosen because they isolate the phenomenon of surface liquid bridging from soaking, dissolution or imbibition that change the dynamics of a granular system.

To this end, the particles are first experimentally characterized using the setups described in the chapter 3. Sensitivities are characterized for their smoothness and experimental data from suitable liquid loading regimes is used to calibrate the numerical model.

4.1. Experimental Characterization

The calibration of the contact model parameters requires determination of a set of particle characteristics, as well as well-defined experimental setups and careful definition of the quantities that the calibration is supposed to be based upon.

The setups have to be chosen in a way that is easily reproducible in simulations and target quantities have to be chosen that are smooth, easily determined in both simulation and experiment, and robust with respect to slight deviations in the setup.

This section is concerned with the experiments that were conducted and their results.

4.1.1. Basic Physical Properties

The system in question are glass particles (Swarco, Germany) with a median diameter $d_{p,50,3} = 1.3\,\mathrm{mm}$, to be used in a wetted state. The physical bulk density was determined by filling a vessel of known volume $V_{vessel} = 817\,\mathrm{cm}^3$ to the brim with particles, resulting in a mass of particles $M_\mathrm{p} = 1.24\,\mathrm{kg}$ and therefore a bulk density of $\rho_\mathrm{p,bulk} = \frac{M_\mathrm{p}}{V_\mathrm{vessel}} = 1520\,\mathrm{kg\,m^{-3}}$.

Wetted particles were produced by placing a known mass of particles into a sampling container, adding desalinated water using a micro-pipette and shaking the particles thoroughly. The final mixing step distributes some of the surface liquid to the surface of the container, but this is assumed to be negligible due to the large surface of the particles ($2.5\,\mathrm{m^2}$ in $1000\,\mathrm{g}$) compared to the surface area of the container ($0.03\,\mathrm{m^2}$ for $5\,\mathrm{cm} \times 5\,\mathrm{cm} \times 12\,\mathrm{cm}$).

4.1.2. Static Angle of Repose

The static angle of repose was measured by filling a cylinder that is capped with a collared, circular base at the bottom, both with a diameter of 10 cm, with particles and removing the collar slowly. The static angles of repose are shown in graphical form in Fig. 4.1 and in Tab. 4.2. The increase in the angle of repose is explained by the acting cohesive forces - the capillary force that acts in the normal direction of particle-particle contacts.

While to be expected, the angle of repose, as determined in this manner, strongly depended on the amount of material put into the cylinder. Shown is the result for $M_0 = 0.4\,\mathrm{kg}$. The result for 0.8 kg is not shown due to a lack of response in the angle of repose. This insensitivity can be explained by the shear behavior during the lifting of the cylinder walls. The stresses that the lower-lying layers of material experience are higher when this larger amount of material flow out as the cylinder is lifted, exceeding the cohesion present due to liquid bridging.

4.1.3. Shear Testing

Shear tests were conducted in a *Schulze Ring Shear Tester RST-XS* with a roughness insert at the bottom lid and a baffled top lid to measure inter-particle friction. For preshearing, a normal stress of $\sigma_\mathrm{preshear} = 5000$ Pa was chosen, as this is on the order of the hydrostatic pressure $p_\mathrm{hydrostatic}$ of tens of centimeters of bulk solid resting:

$$p_\mathrm{hydrostatic} = \rho_\mathrm{bulk} g h = (1520\,\mathrm{kg\,m^{-3}})(9.81\,\mathrm{m\,s^{-2}})(0.3\,\mathrm{m}) \approx 4500\,\mathrm{Pa} \qquad (4.1)$$

where $g = 9.81\,\mathrm{m\,s^{-2}}$ is the specific gravitational acceleration constant and $h = 0.3\,\mathrm{m}$ is the height of material assumed to be relevant to the granulators to be modelled.

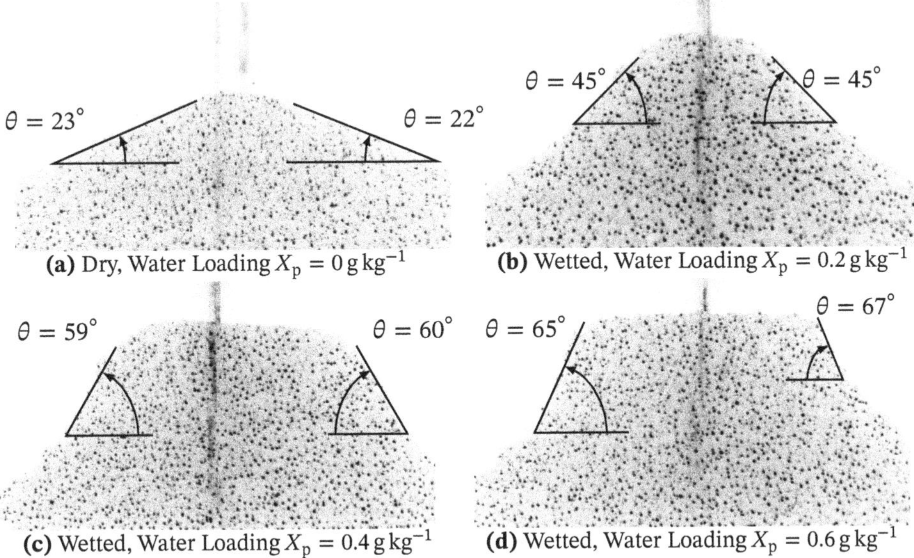

(a) Dry, Water Loading $X_p = 0\,\mathrm{g\,kg^{-1}}$ **(b)** Wetted, Water Loading $X_p = 0.2\,\mathrm{g\,kg^{-1}}$

(c) Wetted, Water Loading $X_p = 0.4\,\mathrm{g\,kg^{-1}}$ **(d)** Wetted, Water Loading $X_p = 0.6\,\mathrm{g\,kg^{-1}}$

Fig. 4.1.: Experimental measurements of the static angle of repose θ of both dry and wetted particles from optical images that have been color-inverted for clarity. No conclusive angle of repose could be determined from $X_p = 0.8\,\mathrm{g\,kg^{-1}}$.

The resulting yield loci are shown in Fig. 4.2 and the result of regressing an equation

$$\tau(\sigma) = \sigma \cdot \tan(\theta) + \tau_c, \tag{4.2}$$

where θ is the internal friction angle and τ_c is the cohesion, is shown in Tab. 4.1.

Tab. 4.1.: Friction angles θ and cohesion τ_c from shear testing yield loci at different water loadings X. Pre-shearing was performed at $\sigma_{\mathrm{preshear}} = 5000\,\mathrm{Pa}$.

Water Loading X [$\mathrm{g\,kg^{-1}}$]	Friction Angle θ [°]	Cohesion τ_c [Pa]
0	29.5	16
0.2	29.8	47
0.4	29.3	123
0.6	27.3	362
0.8	32	432

The shear tests were run four times and the yield locus interpolated for each run, and finally averaged. This allows for the calculation of an estimate for the uncertainty involved with the experiments.

The friction angle (Fig. 4.3a) varies very little between repetitions for each water loading, with standard deviations of measurements for different water loadings overlapping. The cohesion (Fig. 4.3b) shows similar results.

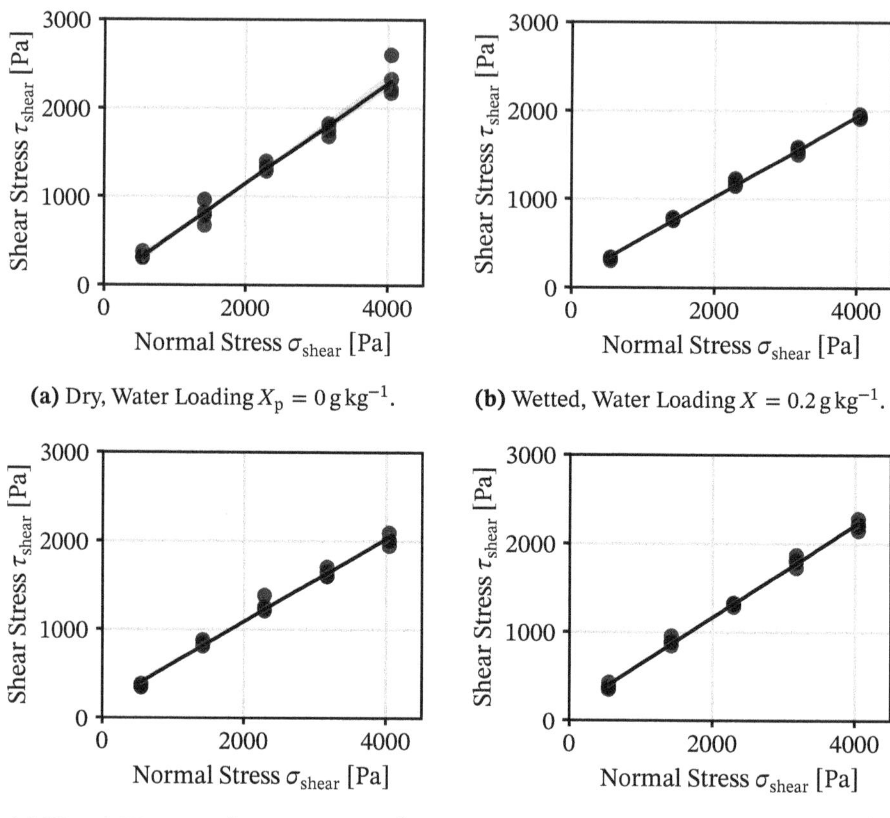

(a) Dry, Water Loading $X_p = 0\,g\,kg^{-1}$.

(b) Wetted, Water Loading $X = 0.2\,g\,kg^{-1}$.

(c) Wetted, Water Loading $X = 0.4\,g\,kg^{-1}$.

(d) Wetted, Water Loading $X = 0.6\,g\,kg^{-1}$.

Fig. 4.2.: Yield loci of the glass particles in both dry and wetted states at a pre-shearing stress of $\sigma_{preshear} = 5000\,Pa$, as well as parameters of the linearized yield loci ($\tau_{shear} = \sigma_{shear}\sin(\theta) + \tau_c$).

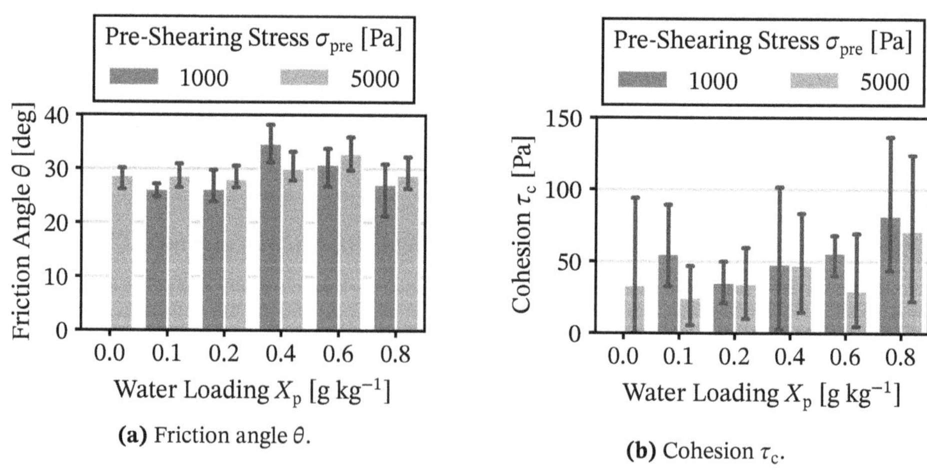

(a) Friction angle θ.

(b) Cohesion τ_c.

Fig. 4.3.: Friction angles θ and cohesion τ_c from shear testing yield loci at different water loadings X.

Therefore, shear testing is not deemed a reliable method for calibrating the liquid bridge model. While commonly known for its reproducibility and accuracy, Richefeu et al. (2006) found a high variability among the Coloumb cohesion when performing shear tests with particle system (monodisperse, $d_p = 1$ mm) similar to the one used in this study. They used a very low (<1 kPa) pre-shearing pressure, unattainable with the equipment present.

4.1.4. Dynamic Angle of Repose

The dynamic angle of repose results are shown in Fig. 4.4 (snapshots),Tab. 4.2 and Fig. 4.5 (plots). The dynamic angle of repose was found to increase with both the rotational velocity and the water loading, although change with respect to the water loading appears to plateau between $0.4\,\mathrm{g\,kg^{-1}}$ and $0.6\,\mathrm{g\,kg^{-1}}$.

Tab. 4.2.: Static and dynamic angle of repose of glass particles ($d_p = 1.3$ mm) wetted with distilled water. A rotational velocity of 0 indicates a static angle of repose experiment. The angle of repose column indicates the mean value and the standard deviation.

Water Loading X [$\mathrm{g\,kg^{-1}}$]	Rotational Velocity ω [$\mathrm{min^{-1}}$]	Angle of Repose Θ [°]
0	0	22.5 ± 0.6
0	12	34.4 ± 2.3
0	25	43.2 ± 0.8
0	60	56.7 ± 3.2
0.2	0	45.0 ± 2.5
0.2	12	39.9 ± 4.3
0.2	25	50.0 ± 0.3
0.2	60	63.1 ± 3.3
0.4	0	59.5 ± 4.8
0.4	12	45.7 ± 1.0
0.4	25	58.1 ± 0.9
0.4	60	69.1 ± 2.0
0.6	0	66.0 ± 1.9
0.6	12	45.2 ± 0.3
0.6	25	61.8 ± 2.5
0.6	60	72.5 ± 1.0
0.8	12	50.0 ± 0.0
0.8	25	59.6 ± 2.3
0.8	60	70.5 ± 2.1

Here, the maximum dynamic angle of repose is around 47° for 12 RPM, 60° for 25 RPM and 70° for 60 RPM. This experiment is very reproducible and gives reliable results, given the low standard deviation of around 2.5° it produces.

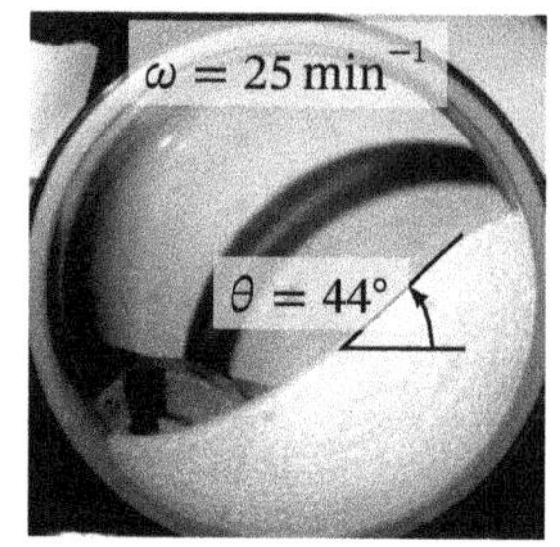

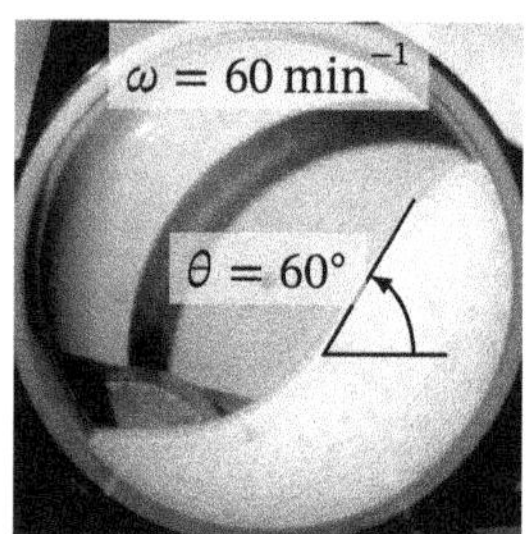

(a) Dry, Water Loading $X_\text{p} = 0\,\text{g}\,\text{kg}^{-1}$

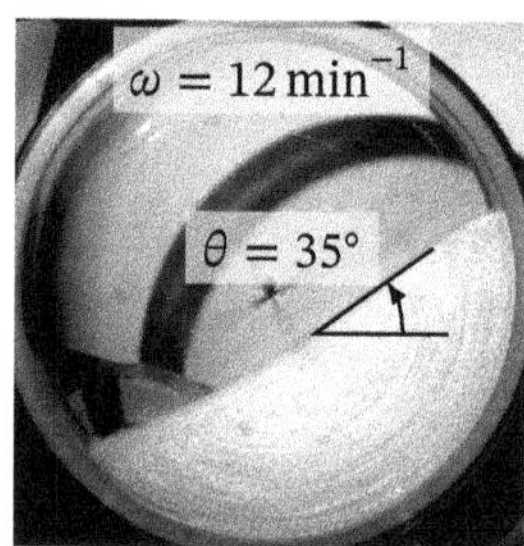

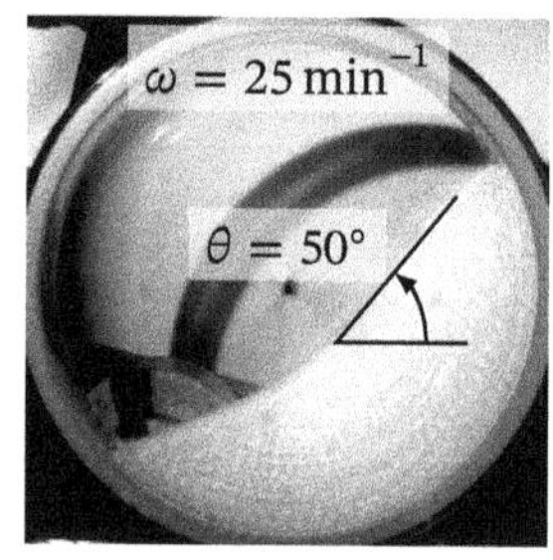

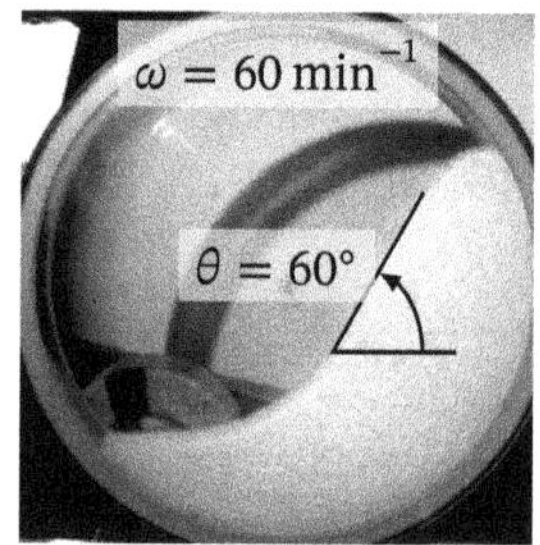

(b) Wetted, Water Loading $X_\text{p} = 0.2\,\text{g}\,\text{kg}^{-1}$

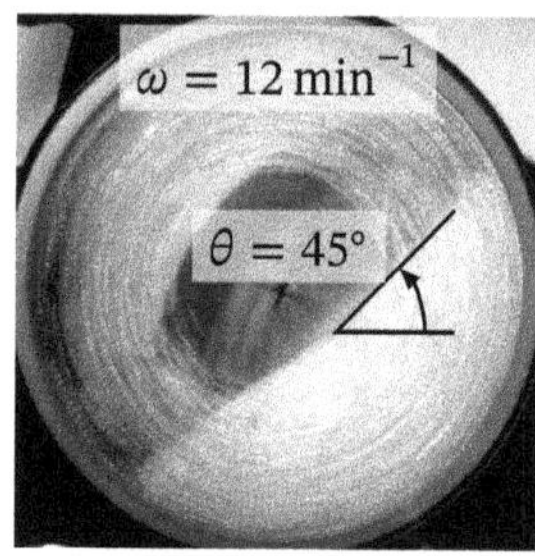

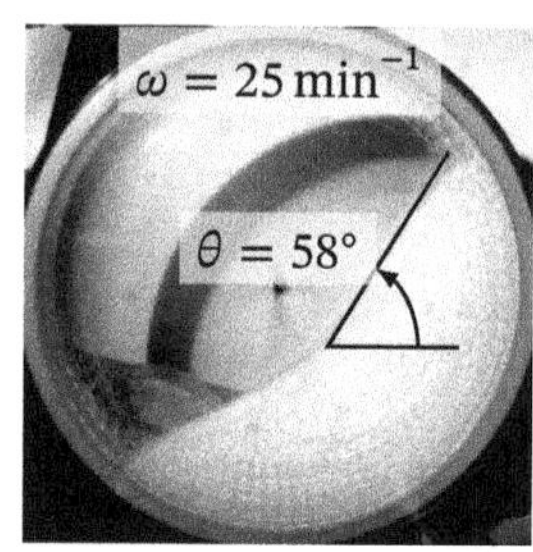

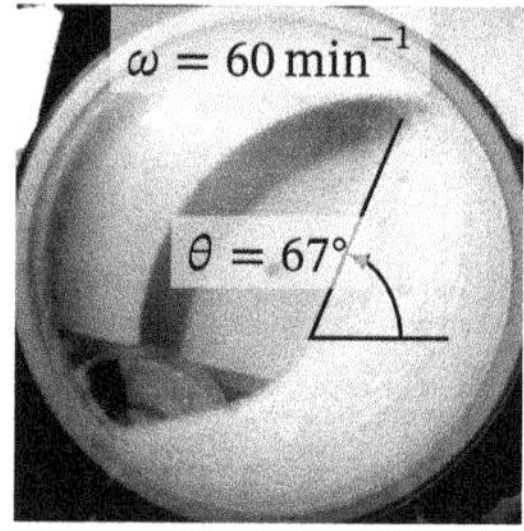

(c) Wetted, Water Loading $X_\text{p} = 0.4\,\text{g}\,\text{kg}^{-1}$

Fig. 4.4.: Determination of the dynamic angle of repose θ of both dry and wetted particles.

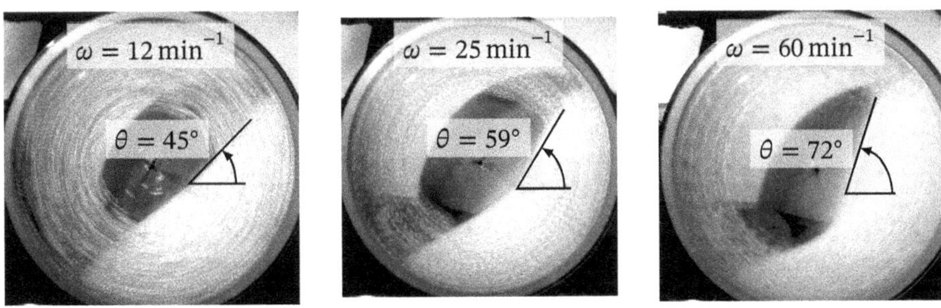

(d) Wetted, Water Loading $X_p = 0.6\,\mathrm{g\,kg}^{-1}$

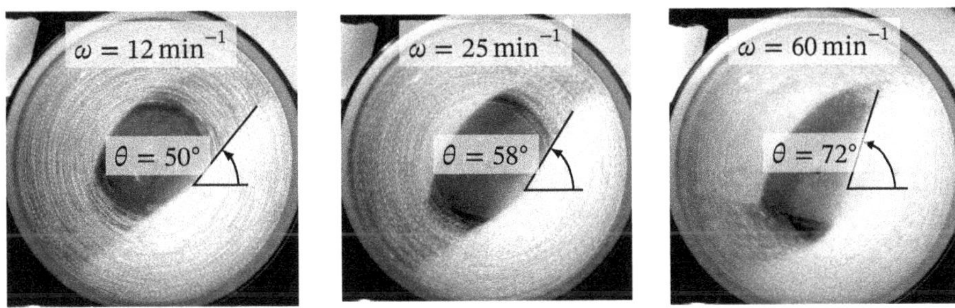

(c) Wetted, Water Loading $X_p = 0.8\,\mathrm{g\,kg}^{-1}$

Fig. 4.4.: Determination of the dynamic angle θ of repose of both dry and wetted particles.

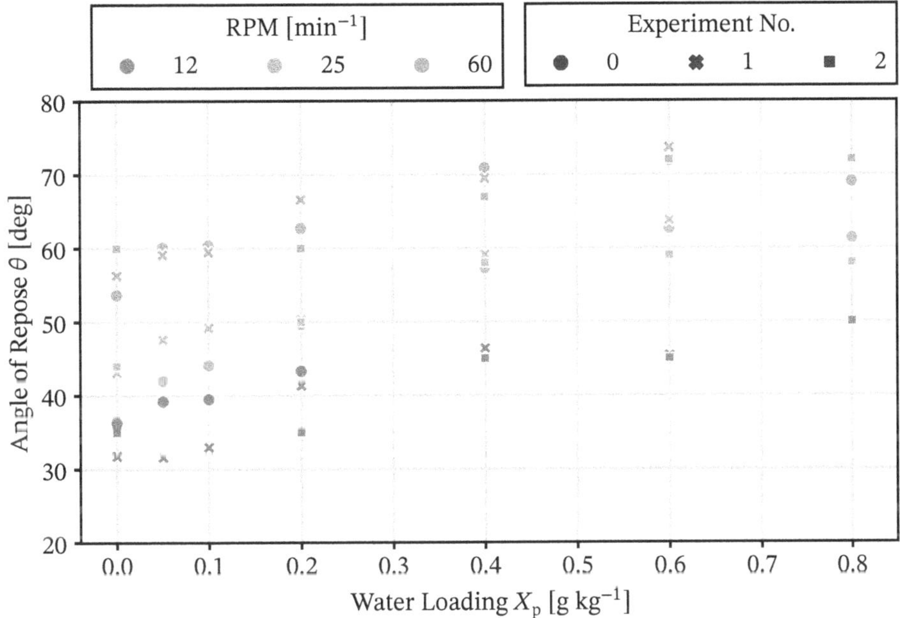

Fig. 4.5.: Dynamic angle of repose test for different water loadings X_p. RPM denotes the drum rotation rate used used.

4.2. Determination of Model Material Parameters

In this section, the key parameters of particle density ρ_p and the particle size distribution $Q_3(d_\mathrm{p})$ are to be determined. As the final ambition of this project is the development of a coarse-grained approach, the particle size distribution is modelled rather than measured.

The particles in question have a narrow particle size distribution, albeit with a low percentage of egg-shaped or plate-shaped contaminants. Using a monomodal size distribution would cause crystalline regions with dense, highly ordered packings that store up energy as they are sheared and release it upon yielding. This was not observable in the shear tests, where small deviations in the particle size distribution and impurities prevent this behavior, as elaborated upon by Wegner et al. (2014). To suppress this behavior, a particle size distribution with three discrete classes was chosen, as shown in Fig. 4.6.

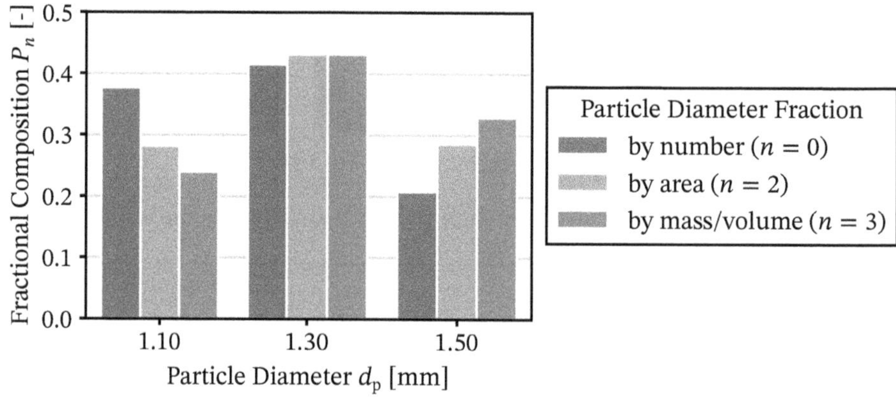

Fig. 4.6.: Particle size distribution used for the calibrations.

The distribution was chosen around a target diameter of 1.3 mm with the two other classes having a distance of 15%. The class weights in the distribution were adjusted to yield a Sauter diameter of $d_{32} = 1.3\,\mathrm{mm}$.

The particle density in the simulations are calibrated by assuming a ballpark-estimate of the solid density and performing a shear test simulation with estimated friction coefficients. The resulting bulk density $\rho_\mathrm{bulk}^\mathrm{sim}$ was then used to scale the assumed solid density $\rho_\mathrm{p}^\mathrm{assumed}$ to match the experimental bulk density $\rho_\mathrm{bulk}^\mathrm{exp}$:

$$\rho_\mathrm{p}^\mathrm{calib} = \rho_\mathrm{p}^\mathrm{assumed} \frac{\rho_\mathrm{bulk}^\mathrm{exp}}{\rho_\mathrm{bulk}^\mathrm{sim}} = \left(2200\,\mathrm{kg\,m^{-3}}\right) \frac{1518\,\mathrm{kg\,m^{-3}}}{1485\,\mathrm{kg\,m^{-3}}} = 2250\,\mathrm{kg\,m^{-3}} \tag{4.3}$$

4.3. Calibration of Frictional Contact Model Coefficients

The frictional parameters for static friction $\mu_{rfr,pp}$ and rolling friction $\mu_{rfr,pp}$ are to be calibrated in this section, based on the dry experimental data given in the previous section.

The static angle of repose test is considered in the calibration using the objective function

$$J_{static}(\mu_{fr,pp}, \mu_{rfr,pp}) = \sqrt{(\Theta_{sim}(\mu_{fr,pp}, \mu_{rfr,pp}) - \Theta_{exp})^2}, \tag{4.4}$$

that is to be minimized by the choice of the friction coefficients that influence the static angle of repose in simulations Θ_{static}. The response surface of the static angle of repose test (Fig. 4.7b) shows a ridge spanning from a parameter set $\{\mu_{fr} = 0.5, \mu_{rfr} = 0\}$ to $\{\mu_{fr} = 0.2, \mu_{rfr} = 0.125\}$, indicating an almost linear contribution from both parameters.

For the shear test, the objective function

$$J_{sheartest}(\mu_{fr,pp}, \mu_{rfr,pp}) = \sqrt{\frac{1}{N_\sigma} \sum_{\text{normal stresses } \sigma} (\tau_{sim}(\sigma, \mu_{fr,pp}, \mu_{rfr,pp}) - \tau_{exp}(\sigma))^2} \tag{4.5}$$

is to be minimized by the choice of the friction coefficients $\mu_{fr,pp}, \mu_{rfr,pp}$ that change the yield shear stress in simulations $\tau_{sim}(\sigma)$ that correspond to a normal stress σ. The resulting response surface in Fig. 4.7a shows the best results can be found for static friction coefficients around 0.35 and high values (> 0.1) for the rolling friction. Analogously, for the dynamic angle of repose an objective function

$$J_{dynamic}(\mu_{fr,pp}, \mu_{rfr,pp}) = \sqrt{\frac{1}{N_\omega} \sum_{\text{rotation rates } \omega} (\Theta_{sim}(\omega, \mu_{fr,pp}, \mu_{rfr,pp}) - \Theta_{exp}(\omega))^2} \tag{4.6}$$

is to be minimized by choosing friction coefficients $\mu_{fr,pp}, \mu_{rfr,pp}$ that influence the angle of repose in the simulations Θ_{sim} at the given rotation rate ω. The objective function (Fig. 4.7c) shows optimal results for high rolling and static friction values, with $\mu_{fr} >= 0.25$ and $\mu_{rfr} >= 0.25$. Interestingly, plotting the objective function for the shear test against that of the drum test (Fig. 4.8a shows a remarkable correlation. This indicates a set of frictional parameters that reduces both $J_{sheartest}$ and $J_{dynamic}$ simultaneously.

On the other hand, comparing the results of the shear test and the static angle of repose (Fig. 4.8b), a front of Pareto-optimal parameters trading off between good reproduction of the shear test and the static angle of repose can be identified.

To reconcile these disparate results, a comprehensive global objective function $J(\mu_{fr,pp}, \mu_{rfr,pp})$ must be defined. As the range of the objective functions at hand are different, defining a trivial linear combination $J = \sum_i(\alpha_i J_i)$ would not suffice.

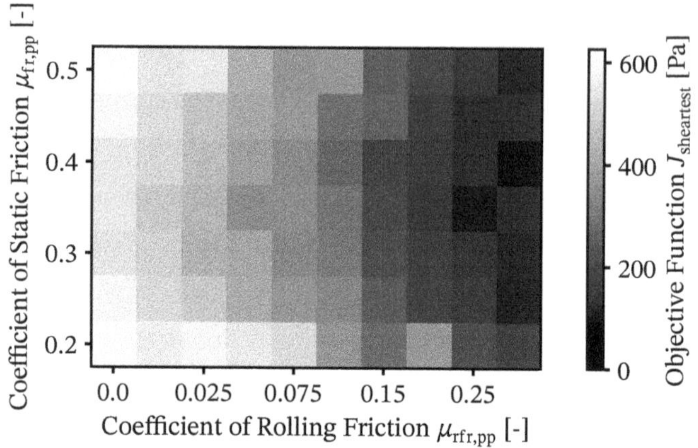

(a) Objective function for the shear test in a dry system.

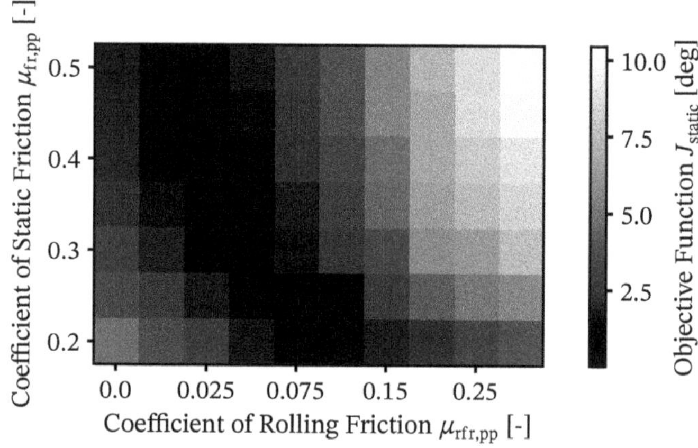

(b) Objective function for the static angle of repose in a dry system.

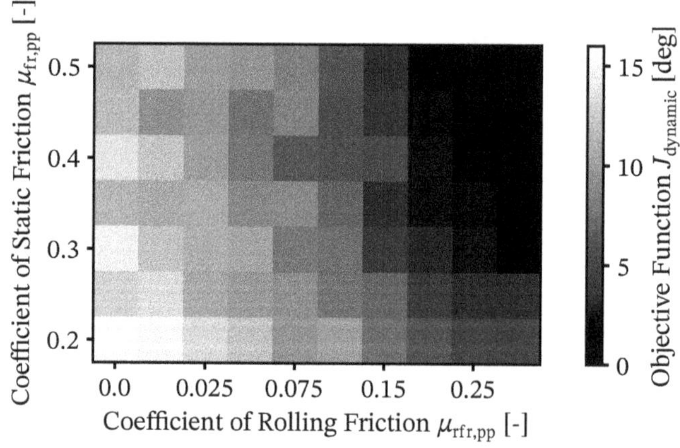

(c) Objective function for the dynamic angle of repose in a dry system.

Fig. 4.7.: Response surface of the objective function for the angle of repose test in a dry system.

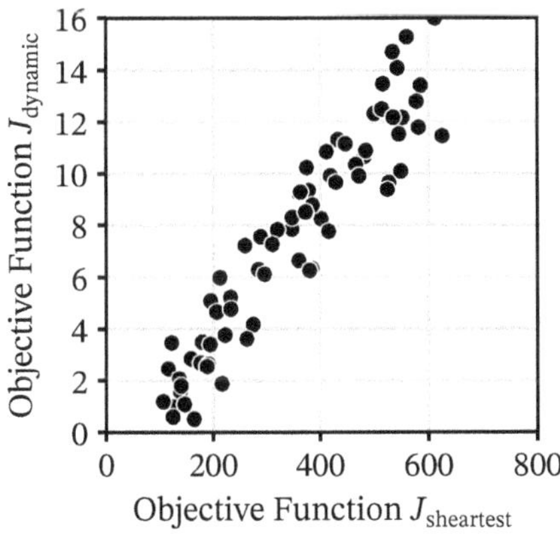

(a) Objective functions of the shear test $J_{sheartest}$ and the dynamic angle of repose $J_{dynamic}$.

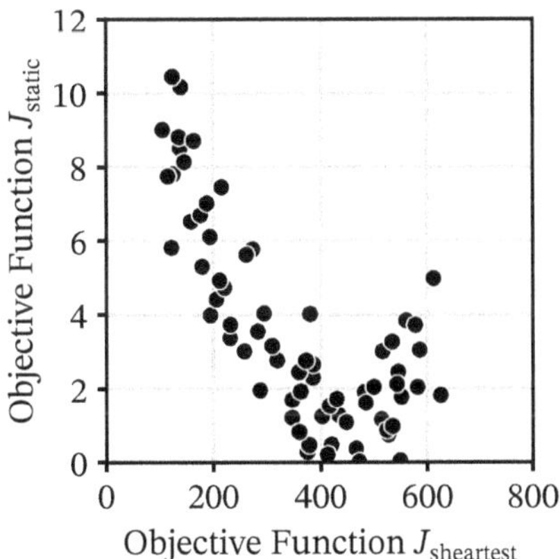

(b) Objective functions of the shear test $J_{sheartest}$ and the static angle of repose J_{static}.

Fig. 4.8.: Plots of the objective functions for the frictional calibration.

Therefore, the values of the objective function are normalized, by subtracting the minimal value observed and dividing by the standard deviation:

$$\{J'_i\} = \left\{\frac{J_i - \min(\{J_i\})}{\mu_1(\{J_i\})}\right\} \tag{4.7}$$

Here, $\{J'_i\}$ is the set of normalized objective function values, $\{J_i\}$ the regular objective function values, $\min(\{J_i\})$ the smallest objective function values in the set and $\mu_1(\{J_i\})$ the standard deviation. Note that this only makes sense after screening a parameter space - otherwise, no guarantees for minimal value or standard deviation can be made. This approach preserves the order within each objective function while normalizing the width of the distribution.

The resulting normalized distributions (Fig. 4.9a) all have values in the range of 0 to 4. Based on this, a new global objective function

$$\begin{aligned} J_{\text{frictional}}(\mu_{\text{fr,pp}}, \mu_{\text{rfr,pp}}) = & \left(J'_{\text{dynamic}}(\mu_{\text{fr,pp}}, \mu_{\text{rfr,pp}})\right)^2 \\ & + \left(J'_{\text{static}}(\mu_{\text{fr,pp}}, \mu_{\text{rfr,pp}})\right)^2 \\ & + \left(J'_{\text{sheartest}}(\mu_{\text{fr,pp}}, \mu_{\text{rfr,pp}})\right)^2 \end{aligned} \tag{4.8}$$

is defined that is the sum of the squares of the individual normalized objective functions. This guarantees that all criteria are included and merely minimizing two will result in large values.

Tab. 4.3.: Frictional parameter sets that minimize the objective function $J_{\text{frictional}}$.

Friction Coefficients		Objective Function			
μ_{fr}	μ_{rfr}	$J_{\text{frictional}}$	J_{dynamic}	$J_{\text{sheartest}}$	J_{static}
0.35	0.15	3.98	3.76	733	4.73
0.3	0.15	4.03	4.77	738	3.73
0.2	0.25	4.06	5.22	735	3.38
0.25	0.25	4.09	3.5	723	5.3
0.2	0.3	4.15	5.08	729	3.99

The issue of finding a Pareto-optimal result that sacrifices one criterion entirely is thus prevented. The distribution of this objective function is shown in Fig. 4.9b where a median of $J_{\text{frictional}} = 5$ can be found, whereas the maximum value is $J_{\text{frictional}} = 23$. As presented in Tab. 4.3, the parameter set $\{\mu_{\text{fr}} = 0.35, \mu_{\text{rfr}} = 0.15\}$ will be used, with an optimal objective function value of $J_{\text{frictional}} = 3.98$. The resulting deviation from the dynamic angle of repose is less than 2 ° at any rate of rotation and the static angle of repose matches within 1 °.

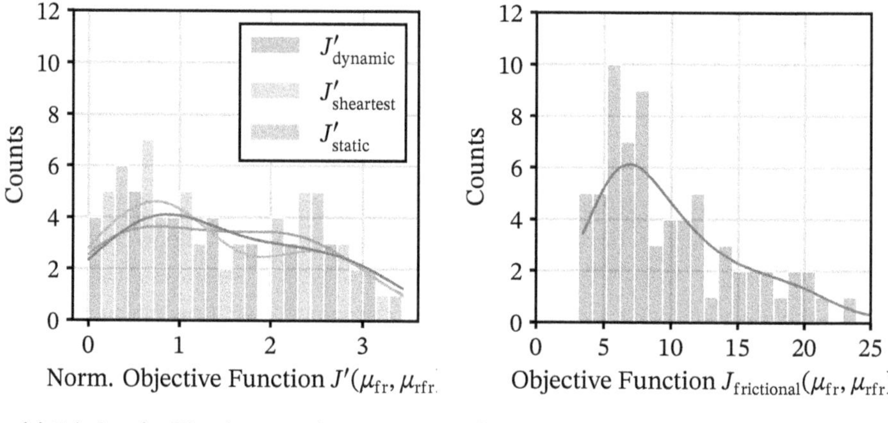

(a) Frictional calibration experiments. **(b)** Overall frictional objective function.

Fig. 4.9.: Distributions of the normalized objective function for frictional calibration for each of the calibration experiments (Fig. 4.9a) and the overall objective function for frictional calibration (Fig. 4.9b), and their respective smoothened approximations.

4.4. Calibration of Liquid Bridge Model Coefficients

The liquid bridge model minimum layer height h_{min} and particle minimum separation distance $\delta_{lb,min}$ have a large influence on inter-particle contacts, as elaborated previously (section 3.4.4).

The minimum liquid layer height h_{min} can be principally calibrated using the static angle of repose test in an isolated manner, drastically reducing the parameter space to be searched. The objective function is formulated in analogy to (4.4), but considers the deviations at all water loadings X:

$$J^{\text{wetted}}_{\text{static}}(h_{\min}, \sigma_{\text{surface}}) = \sqrt{\frac{1}{N_X} \sum_{\text{water loading } X} \left(\theta_{\text{sim}}(h_{\min}, \sigma_{\text{surface}}, X) - \theta_{\text{exp}}(X)\right)^2} \tag{4.9}$$

The mean angle of repose value is used for the experiments and the sum is extended to average over the different sections of the simulation (see procedure).

After determining the parameter space of h_{min} to be bounded by a value of $0.08\,\mu\text{m} \leq h_{min} \leq 1\,\mu\text{m}$ by using the surface tension value of water, $\sigma_{surface} = 0.073\,\text{N}\,\text{m}^{-2}$, the surface tension range of $0.05\,\text{N}\,\text{m}^{-2} \leq \sigma_{surface} \leq 0.2\,\text{N}\,\text{m}^{-2}$ was explored for $h_{min} = 0.4\,\mu\text{m}$ and $h_{min} = 1\,\mu\text{m}$ to see if the difference in number in liquid bridges created for high values of h_{min} could be compensated by bridge strength.

The resulting distribution of objective function values is given in Fig. 4.10a. While for the objective functions of the static angle of repose J'_{static} center around optimality, this is not the case for the objective functions of the static angle of repose $J'_{dynamic}$. Here, the values cluster around a value of 3.5. This indicates that weak optimality can be reached with more than one parameter combination and the parameter space was chosen unnecessarily large. The Pareto plot in Fig. 4.10b confirms this conjecture.

The best performing parameter sets from this study that minimize the aforementioned objective function are given in Tab. 4.4. Among these were primarily those using the original surface tension and varying values of the minimum layer height h_{min}. Notably, the best parameter set is one that uses the upper bound of the minimum layer height range with a surface tension that is higher than the physical value.

After the static liquid bridge parameters, those influencing dynamic behavior, namely the minimum separation height $\delta_{lb,min}$ and the viscosity μ_l, have to be calibrated. Due to the exploratory nature of this calibration run, both surface tension and minimum layer height were varied as well, but this is not strictly necessary.

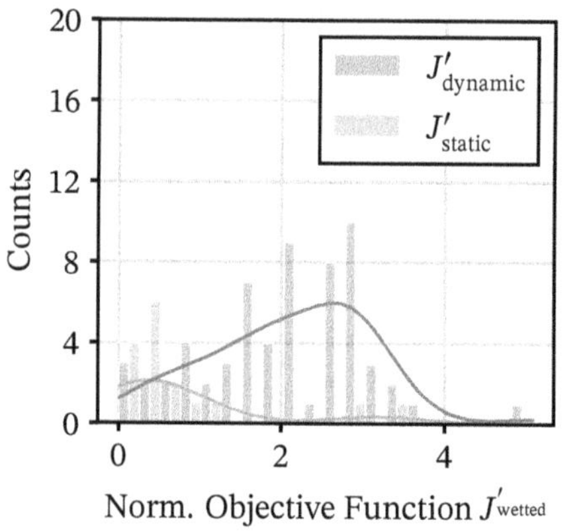

(a) Liquid bridge model calibration experiments.

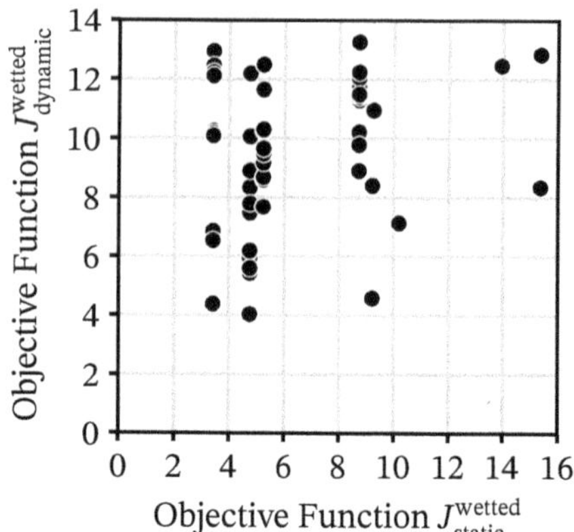

(b) Pareto plot of the objective functions of the wetted calibration experiments.

Fig. 4.10.: Distributions of the normalized objective function for liquid bridge model calibration for both of the calibration experiments (Fig. 4.10a), and their respective smoothened approximations, and the corresponding Pareto plot (Fig. 4.10b).

Tab. 4.4.: Liquid bridge parameter sets that minimize the objective function J^{wetted}_{static} of the static angle of repose test.

Rank	Score J^{wetted}_{static}	Min. Height h_{min} [µm]	Surface Tension $\sigma_{surface}$ [N m^{-2}]
1	3.42	1	0.1
2	4.15	0.8	0.073
3	4.75	0.4	0.073
4	5.23	0.6	0.073
5	7.77	0.08	0.073

The dynamic angle of repose test (Tab. 4.5) confirms the validity of the choice of angle of repose: The two best-ranking dynamic angle of repose runs agree on the values for minimum layer height and surface tension. The combinations of $h_{min} = 0.4$ µm, $\sigma_{surface} = 0.0728$ N m^{-1} and $h_{min} = 0.4$ µm, $\sigma_{surface} = 0.0728$ N m^{-1} are the two best performing parameter sets that trade off between liquid bridge connectivity (relating to h_{min}) and static liquid bridge strength ($\propto \sigma_{surface}$). Analogously, rank two and three as well as two anti-correlating parameters determine the degree of viscous dissipation ($\delta_{lb,min}$ - ν_l). The best parameter set uses a physical value for the surface tension, but not for the liquid phase viscosity. This being selected in accordance with a very low value for the minimum separation distance indicates that by screening the minimum separation distance in more detail.

Tab. 4.5.: Liquid bridge parameter sets that minimize the objective function $J^{\text{wetted}}_{\text{dynamic}}$ of the dynamic angle of repose test.

Rank	Score $J^{\text{wetted}}_{\text{dynamic}}$	Min. Height h_{min} [µm]	Surface Tension σ_{surface} [N m^{-1}]	Min. Sep. $\delta_{\text{lb,min}}$ [µm]	Viscosity ν_{l} [mPa s]
1	4.03	0.4	0.073	6.5	0.1
2	4.38	1	0.1	656.5	0.89
3	4.58	0.4	0.1	656.5	0.89
4	5.23	0.2	0.1	656.5	0.89
5	5.43	0.4	0.073	6.5	0.75
6	5.59	0.4	0.073	65	0.89
7	5.94	0.4	0.073	32.5	0.89
8	6.18	0.4	0.073	6.5	1
9	6.28	1	0.15	6.5	0.89
10	6.52	1	0.1	656.5	0.75

4.5. Summary

A workflow for calibrating liquid bridge laws in fine detail was demonstrated. The workflow's segregated nature was used to confirm the choice of - at least - the static contact model parameters. The large parameter space was reduced using Pareto analysis and efficient objective functions for optimization were formulated. The interplay between the correlating parameter pairs, namely the minimum layer height-surface tension and minimum separation distance-liquid viscosity, was shown and a set of applicable contact model parameters were presented.

5 Implementation and Validation of a Full-Physics CFD-DEM Solver for the Simulation of Fluidized Bed Spray Granulation

This chapter details the development of a solver capable of describing the drying of surface liquid films, commonly called the first drying phase, in particle-gas flows based on the CFDEMcoupling framework. The solver can also describe the wetting process due to the flow of droplets in the gas phase. This will provide the capability to simulate the drying performance of fluidized and spouted beds, as well as provide a solid foundation for further work that correlate tracked quantities with the formation of surface structures during granulation. The microscopic processes that were implemented are

- droplet injection and transport in the gas phase,
- droplet evaporation,
- deposition of droplets onto the particle surface,
- evaporation of liquid on the particle surface,
- transport of energy/enthalpy in the gas phase and
- transport of a vapor species in the gas phase.

A focus was laid on the modeling of surface liquid, as it greatly influences the accuracy of predicted drying and heat transfer rates that are to be correlated with the surface structures. Energy and mass conservation were verified to ensure correctness of the implementation (see Appendix A).

The solver, named *cfdemSolverEvap* is based on two solvers, *cfdemSolverRhoPImpleChem* and *sprayFoam*. The first, *cfdemSolverRhoPImpleChem* was originally developed by Schneiderbauer et al. (2020) and is a compressible CFD-DEM solver that includes support for implicit heat transfer and species transport with an explicit source term. *sprayFoam* is contained in OpenFOAM for the purpose of combustion modeling and, as such, contains models for evaporation and heat transfer that are leveraged in this code.

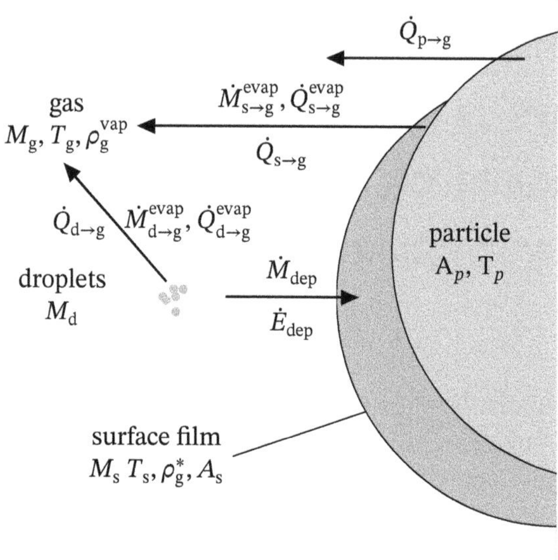

Fig. 5.1.: Overview of the control volumes and heat and energy fluxes.

First, the governing equations for the heat and mass balances of surface liquid on individual particles are presented. The subsequent section aims to give an overview of the different surface coverage models and their applicability to coupling the first-phase drying rate with material behavior.

Finally, the governing equations for vapor and heat transport and their numerical treatment are shown.

5.1. Governing Equations for Heat and Mass Transfer Modeling

This chapter will provide insight in the governing equations used in the developed solver, and subsequently explain and justify the choices made. An illustration of the phase interactions is given in Fig. 5.1.

5.1.1. Governing Equations for the Droplet Phase and the Particle Phase

In this section, the source terms for the modeling of evaporation for the discrete Lagrangian elements in the simulation are given.

Evaporation occurs in the context of four different control volumes, namely the gas phase (g), the solid particle phase (p) and the surface film (s) on the surface of said particles, as well as the droplets (d). The focus is therefore on their interactions driven by evaporation and heat exchange, and formulating their differential heat and mass balances. Finally, a set of suitable closures to these balances is presented.

The vapor mass source term for the gaseous phase is given by

$$\frac{dM_{\mathrm{g}}}{dt} = \sum_i \dot{M}^{\mathrm{evap}}_{\mathrm{s}\to\mathrm{g},i} + \sum_j \dot{M}^{\mathrm{evap}}_{\mathrm{d}\to\mathrm{g},j} \tag{5.1}$$

where M_{g} refers to the mass of vapor in a control volume, $\dot{M}^{\mathrm{evap}}_{\mathrm{s}\to\mathrm{g},i}$ is the vapor mass flow emitted by the surface liquid of a single particle i into the gas phase and $\dot{M}^{\mathrm{evap}}_{\mathrm{d}\to\mathrm{g},j}$ is the respective vapor mass flow emitted by a single droplet j into the gas phase.

The corresponding change in internal energy or enthalpy due to the process of evaporation, referred to summarily by E_{g} or energy, is given by

$$\frac{dE_{\mathrm{g}}}{dt} = \sum_i (\dot{E}^{\mathrm{evap}}_{\mathrm{s}\to\mathrm{g},i} + \dot{Q}_{\mathrm{p}\to\mathrm{g},i} + \dot{Q}_{\mathrm{s}\to\mathrm{g},i}) + \sum_j (\dot{E}^{\mathrm{evap}}_{\mathrm{d}\to\mathrm{g},j} + \dot{Q}_{\mathrm{d}\to\mathrm{g},j}), \tag{5.2}$$

where for any given particle i, the enthalpy flow to the gas phase due to evaporation is given by $\dot{E}^{\mathrm{evap}}_{\mathrm{s}\to\mathrm{g},i}$, the heat exchanged with the surface film and the unwetted particle surface by $\dot{Q}_{\mathrm{s}\to\mathrm{g},i}$ and $\dot{Q}_{\mathrm{p}\to\mathrm{g},i}$, respectively. Analogously, $\dot{E}^{\mathrm{evap}}_{\mathrm{s}\to\mathrm{g},i}$ represents the energy flow due to evaporation from a droplet j and $\dot{Q}_{\mathrm{d}\to\mathrm{g},j}$ the heat exchanged with said droplet.

The surface film is located on every individual particle and possesses both a mass $M_{\mathrm{s},i}$ and an energy $E_{\mathrm{s},i}$. The change in surface liquid mass over time is given by

$$\frac{dM_{\mathrm{s},i}}{dt} = \dot{M}^{\mathrm{dep}}_i - \dot{M}^{\mathrm{evap}}_{\mathrm{s}\to\mathrm{g},i}, \tag{5.3}$$

where $\dot{M}^{\mathrm{dep}}_i$ is the mass flow rate due to deposition from droplets.

The corresponding energy balance is

$$\frac{dE_{\mathrm{s},i}}{dt} = \dot{E}^{\mathrm{dep}}_i - \dot{E}^{\mathrm{evap}}_{\mathrm{s}\to\mathrm{g},i} - \dot{Q}_{\mathrm{s}\to\mathrm{g},i} - \dot{Q}_{\mathrm{s}\to\mathrm{p},i}, \tag{5.4}$$

where $\dot{E}^{\mathrm{dep}}_i$ is the energy flow due to deposition of droplets, and $\dot{Q}_{\mathrm{s}\to\mathrm{p},i}$ is the heat exchanged with the particle.

For the particle, only the change in energy $E_{p,i}$ has to be considered:

$$\frac{dE_{p,i}}{dt} = \dot{Q}_{s\to p,i} - \dot{Q}_{p\to g,i}. \tag{5.5}$$

5.2. Closures for Heat and Mass Transfer

5.2.1. Heat Transfer

Heat transfer is described using the Gunn correlation (Gunn, 1978) for the Nusselt number:

$$\begin{aligned} \mathrm{Nu} = \frac{\alpha d_p}{k_g} =& (7 - 10\alpha_g + 5\alpha_g^2)(1 + 0.7\mathrm{Re}^{0.2}\mathrm{Pr}^{0.33}) \\ &+ (1.33 - 2.4\alpha_g + 1.2\alpha_g^2)\mathrm{Re}^{0.7}\mathrm{Pr}^{0.33}, \end{aligned} \tag{5.6}$$

where $\mathrm{Re} = U_{rel}\rho_g/\eta_g$ is the Reynolds number, d_p is the particle diameter, U_{rel} is the relative velocity between particle and gas phase, ρ_g is the gas density and η_g is the gas phase viscosity, α is the heat transfer coefficient, k_g is the heat conductivity of the gas, $\mathrm{Pr} = c_{p,g}\eta_g/k_g$ is the Prandtl number, $c_{p,g}$ is the heat capacity of the gas phase and α_g the volume fraction of gas in the proximity of the particle. This correlation captures the swarm effect of other particles surrounding the particle considered for Reynolds numbers $\mathrm{Re} < 10^5$ and solids volume fractions $\alpha_p > 0.35$.

The resulting heat flow between particle i and gas is given by:

$$\dot{Q}_{g\to p,i} = \frac{\mathrm{Nu} d_p}{k_g} S(1 - \varphi_{wet,i})(T_g - T_{p,i}), \tag{5.7}$$

where T_g is the gas temperature at the location of particle i and $T_{p,i}$ is the particle temperature.

Consequently, the heat flow of the surface liquid on particle i and the gas amounts to

$$\dot{Q}_{g\to s,i} = \frac{\mathrm{Nu} d_p}{k_g} S_p \varphi_{wet,i}(T_g - T_{s,i}), \tag{5.8}$$

where $T_{s,i}$ is the surface liquid temperature.

As particle and liquid film are in contact, heat transfer between them occurs as well. It is assumed that the film is stagnant, thus the limit of Nu = 2 can be used:

$$\dot{Q}_{\mathrm{p}\to\mathrm{s},i} = \frac{2h_{\mathrm{film}}}{k_{\mathrm{liq}}} S_{\mathrm{p}} \varphi_{\mathrm{wet},i} (T_{\mathrm{p},i} - T_{\mathrm{s},i}), \tag{5.9}$$

where k_{liq} is the thermal conductivity of the surface liquid.

5.2.2. Mass Transfer

The evaporation rate $\dot{M}^{\mathrm{evap}}_{\mathrm{s}\to\mathrm{g},i}$ can be calculated using the analogy between heat and mass transfer. The analogy states that a heat transfer correlation, such as that given in Eq. (5.6) by Gunn (1978) can be used for the calculation of the mass transfer coefficient β by substituting the Nusselt number Nu with the Sherwood number

$$\mathrm{Sh} = \frac{\beta D}{d_{\mathrm{p}}}, \tag{5.10}$$

where D is the diffusion coefficient of the vapor species in the bulk and the Prandtl number Pr with the Schmidt number $\mathrm{Sc} = \eta_{\mathrm{g}}/\rho_{\mathrm{g}} D$. From this, the evaporation rate of an individual particle can be calculated using:

$$\dot{M}^{\mathrm{evap}}_{\mathrm{s}\to\mathrm{g},i} = \frac{\mathrm{Sh} d_{\mathrm{p}}}{D} S_{\mathrm{s},i} \varphi_{\mathrm{wet},i} \rho_{\mathrm{g}} (y_i^* - y_i), \tag{5.11}$$

where $S_{\mathrm{p}} = \pi d_{\mathrm{p}}^2$ is the total particle surface area and $\varphi_{\mathrm{wet},i}$ the fraction of said area wetted by the liquid film. y is the gas phase vapor fraction and y_i^* the saturation vapor fraction in the gas phase, derived from an Antoine equation:

$$y_i^* = \frac{1 \cdot 10^5\,\mathrm{Pa\,bar^{-1}}}{\rho_{\mathrm{g}} R_{\mathrm{m}} M_{\mathrm{w,vap}} T_{\mathrm{s},i}} 10^{A - \frac{B}{T_{\mathrm{s},i} + C}}, \tag{5.12}$$

where $R_{\mathrm{m}} = 8.314\,\mathrm{J\,mol^{-1}\,K^{-1}}$ is the molar gas constant, M_{vap} is the molecular weight of the vapor species and $T_{\mathrm{s},i}$ the surface film temperature. For water, the Antoine equation coefficients are given by $A = 4.6543$, $B = 1435.264\,\mathrm{K}$, $C = -64.848\,\mathrm{K}$, and valid in the range of $255.9\,\mathrm{K} \leq T \leq 373\,\mathrm{K}$. the molecular weight M_{vap} assumes a value of $0.018\,02\,\mathrm{kg\,mol^{-1}}$, as given by Stull (1947).

5.3. Surface Coverage Modeling

As outlined in the previous section, the covered surface fraction φ_{wet} plays a central role in evaporation modeling. Before implementing this, its influence on simple drying scenarios will be analyzed. An overview over the different ways in which the liquid content on a particle's surface can be treated for the purpose of depicting evaporation is given in Fig. 5.2.

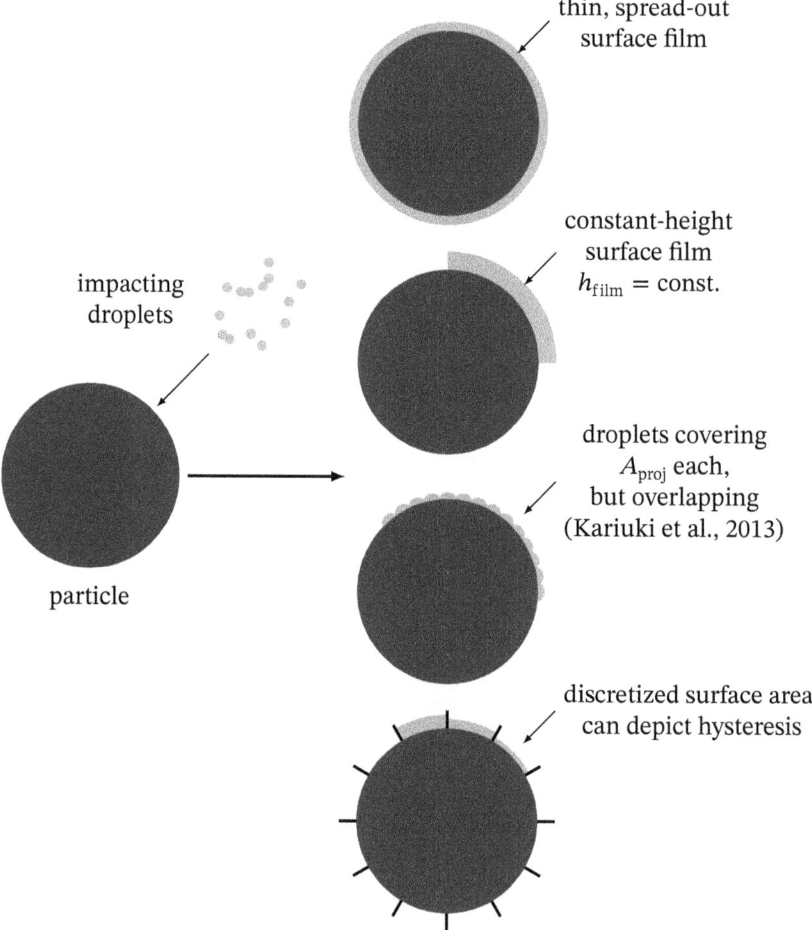

Fig. 5.2.: Schematic of the different ways of treating impacting droplets.

5.3.1. Analytical Surface Coverage Models

Analytical surface coverage models only depend on the mass of liquid present on the surface, as well as geometrical parameters. This implies that hysteresis and history effects cannot be captured. The hysteresis and history effects influence drying times directly, and thus the kinetics of surface structure-forming in granulation (see chapter 6).

Numerically, analytical models are very cheap both in terms of computing power and memory demand. The latter is especially important in DEM simulations, as the entirety of the particles' state is transferred during processor crossings.

The first, trivial model is that of *instantaneous, complete wetting*, depicted at the top of Fig. 5.2. It assumes that the presence of even one droplet immediately covers the entire surface. Mathematically, it can be formulated as

$$\varphi_{\mathrm{wet},i} = \begin{cases} 1 & M_i > 0 \\ 0 & \text{else.} \end{cases} \tag{5.13}$$

This model may be applicable for systems with very small contact angles and low viscosity where the surface liquid spreads immediately.

Alternatively, one may use a *constant film height*, where the liquid is assumed form a film of a certain height on a partially wetted surface:

$$\varphi_{\mathrm{wet},i} = \min\left(\frac{M_{\mathrm{l},i}}{M^*}, 1\right), \tag{5.14}$$

where M^* is the mass of liquid that would result in a film of thickness h:

$$M^* = \rho_{\mathrm{l}} \frac{\pi}{6}\left((d_{\mathrm{p}} + 2h)^3 - d_{\mathrm{p}}^3\right) \tag{5.15}$$

This shows the behavior of a liquid that has a high surface tension and contact angle, i.e. a liquid that reduces the surfaces it exposes, both to the gas phase and the particle surface.

The model by Kariuki et al. (2013) assumes droplets to wet their projection area $A_{\mathrm{proj}} = \pi d_{\mathrm{d}}^2/4$, equating to an individual fractional coverage of $f = A_{\mathrm{proj}}/S_{\mathrm{p}} = (d_{\mathrm{d}}/2d_{\mathrm{p}})^2$. The areas of successive impacts are assumed to overlap with previous impacts with a probability of $(1-f)^{(N-1)}$, with $N-1$ being the number of previous impacts. Thus, the wetted area fraction approaches a value of

$$\varphi_{\mathrm{wet},i} = 1 - (1-f)^N, \tag{5.16}$$

This has to be generalized to non-integer values of N by approximating the number of impacts as the ratio of surface liquid mass M_i and the droplet mass $M_\mathrm{d} = \rho_\mathrm{d}\pi d_\mathrm{d}/6$:

$$N = M_i/M_\mathrm{d} \tag{5.17}$$

This introduces inaccuracies in the case of polydisperse sprays, as either the mean diameter of impacted droplets would have to be tracked, or would need to be estimated *a priori*.

5.3.2. Surface Coverage Modeling using a Discretized Surface Area

For evaluating the benefits of a surface-discretized approach, one such model was implemented. The goal of this was to identify the general properties of such discretizations. Instead of using a tessellation, i.e. from an STL file, a surface is described by N equal areas $S_{\mathrm{s},i}$ with a circular topology, with each of the areas tracking the collected mass M_i, given by:

$$S_{\mathrm{s},i} = \frac{S}{N}, \quad i = 1 \dots N. \tag{5.18}$$

Spray deposition was modeled by randomly selecting one of the areas and distributing the droplet among neighboring areas based on the ratio of droplet cross section and area $1/f$:

$$1/f = \frac{S}{NA_{\mathrm{proj,d}}}, \tag{5.19}$$

$$M_j \mathrel{+}= \frac{M_d}{1/f}, \quad j = \{i - 1/f/2, \dots i, \dots i + 1/f/2\} \tag{5.20}$$

Areas contribute all of their area to the wetted surface area if it contains any liquid mass:

$$\varphi_\mathrm{wet} = \frac{\sum_{j=1}^{N}((M_j > 0)\,?\,1\ :\ 0)}{N}, \tag{5.21}$$

Where ? : is the ternary operator. During drying, liquid mass is removed from all wetted areas in equal parts, as the mass rate is proportional to the area wetted.

5.3.3. Comparison of Surface Coverage Models

The previously described models were implemented and tested using the following exemplary drying cycle: first, droplets with a diameter of $d_\mathrm{d} = 100\,\mu\mathrm{m}$ were deposited on a particle with a diameter of $d_\mathrm{p} = 2000\,\mu\mathrm{m}$ at a rate of 1 per time step for 1000 steps. Then, drying using a rate of

$$\dot{M} = \varphi_\mathrm{wet}\frac{M_\mathrm{d}}{10} \tag{5.22}$$

equating to a tenth of a droplet mass per time step was performed for 10 000 steps. In all cases, the particle was dried at least partially. This was repeated with another wetting cycle of 10 000 steps to cause the surface coverage to approach 100% and a final drying cycle of 240 000 steps to check for hysteresis effects.

The resulting wetted surface fraction and liquid mass over time is shown in Fig. 5.3.

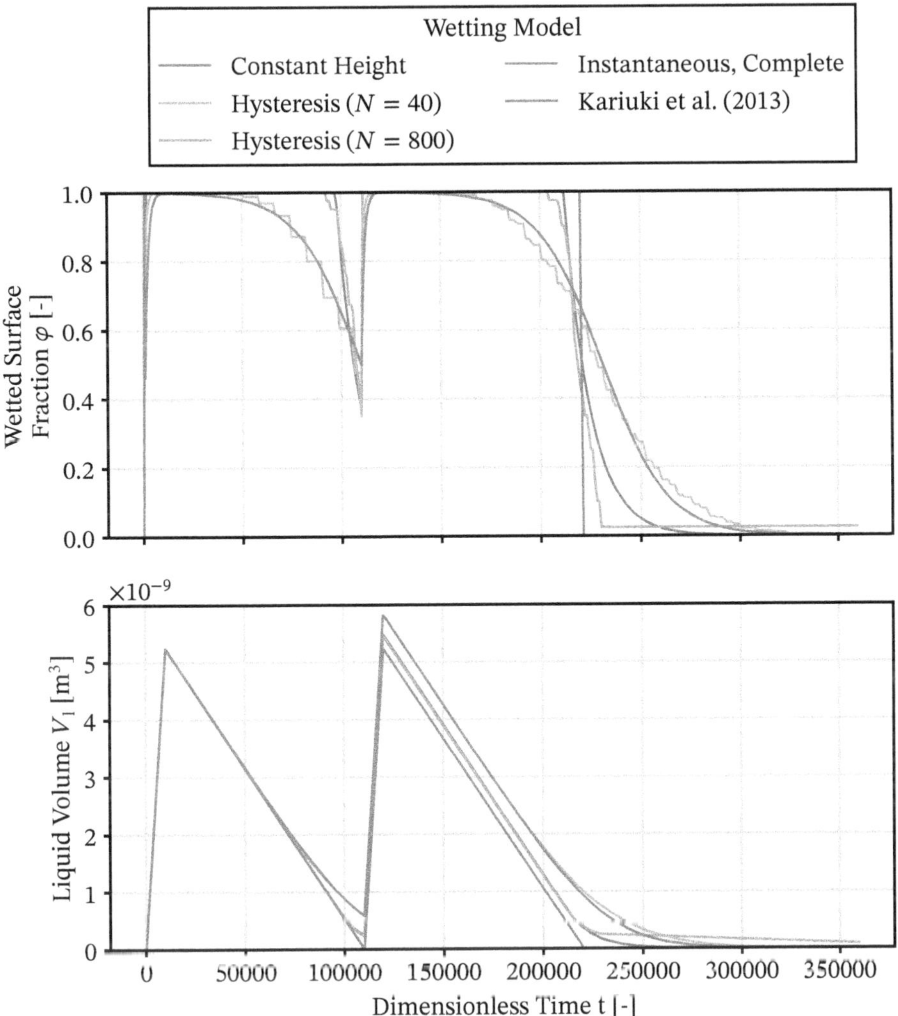

Fig. 5.3.: Wetted surface fraction and surface liquid mass for two wetting-drying cycles using different surface coverage models.

As expected, the instant, total wetting model predicts the fastest drying. Of the non-trivial, analytical models, the minimum layer height model predicts the fastest drying, although this can be adjusted using the layer height, arbitrarily chosen to be $h = d_d/2$ to include the influence of slight spreading in this example. The Kariuki et al. (2013) approach, configured to use the actual droplet and particle sizes, shows a much slower drying behavior. This is due to the *saturating* nature of the model in the case of higher wetting, meaning that the wetted area decreases much earlier.

Two discretizations as described in the previous subsection were included: Hysteresis ($N = 40$) and hysteresis ($N = 800$), differing only in the number of discretized areas N.

One can easily see that the two analytical models are border cases of area discretizations - the finer discretization hysteresis ($N = 800$) approximates the minimum layer height model and the coarser hysteresis ($N = 40$) approaches the Kariuki et al. (2013) model. This is due to the stochastical foundations of both the minimum layer height and the Kariuki et al. (2013) model.

In the second drying cycle, one can observe the hysteresis effect captured by discretizations and its influence on the drying behavior. Although both discretizations approximated their analytical counterparts perfectly in the first cycle, their drying rate during the second cycle was much slower. This is caused by hysteresis effects not captured by the analytical models.

5.4. Droplet Deposition Modeling

While droplet deposition onto particles is most intuitively modeled using direct impact calculation, this approach is numerically expensive. Instead, droplet deposition is modeled using a filter correlation. This avoids the need for contact detection, but introduces further assumptions. Previous works (Kieckhefen et al., 2018a) employed a deposition model that uses a filter correlation by Kolakaluri (2013) to calculate a collection efficiency, as outlined in Fig. 5.4.

The CFD grid is used as a set of control volumes for droplet deposition. The particle volume fraction is already known and the association between the particle/droplet and the cell already exists, lowering the numerical cost of the method. For each cell, the Sauter diameter $d_{32} = \frac{\sum_j d_j^3}{\sum_j d_j^2}$ of both droplets and particles is calculated, as well as the surface fraction of each particle among all particles present. The relative velocity U_r is calculated from the mass-averaged velocity of the droplets relative to the mean particle velocity. In each coupling step, the droplet parcels are stripped of individual parcels and the mass distributed among all particles in the cell. The fraction of droplets in a parcel deposited is calculated using the set of equations given in Tab. 5.1.

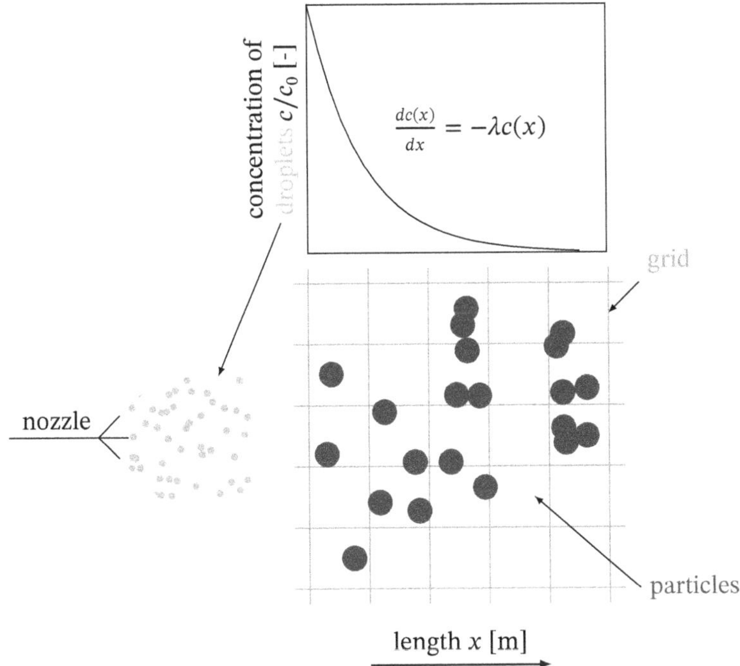

Fig. 5.4.: Schematic of the filter model used to describe the deposition process of droplets onto the particles.

Tab. 5.1.: Model quantities of the Kolakaluri (2013) filter deposition model.

Quantity	Expression	Unit
Mean Reynolds number	$\mathrm{Re_m} = \alpha_g U_r d_{p,32}/\eta_g$	–
Stokes number	$\mathrm{St} = U_r d_{d,32}^2 \rho_d / 18 d_{p,32} \eta_g$	–
Model parameter	$A = \frac{1-\alpha_p^{5/3}}{1-1.5\alpha_p^{1/3}+1.5\alpha_p^{5/3}-\alpha_p^2}$	–
Effective Stokes Number	$\mathrm{St_{eff}} = 0.5\,\mathrm{St}\left(A + 1.14\mathrm{Re}^{1/5}\alpha_g^{-3/2}\right)$	–
Filter coefficient	$\lambda = \mathrm{St_{eff}^{3.2}}/(4.3 + \mathrm{St}^{3.2})$	m^{-1}
Filter efficiency	$\eta_{coll} = 1.5\lambda U_r \alpha_p / d_p$	–

This model does not directly capture the effect of polydispersity with regard to spray or particles but instead it assumes that the Sauter diameter is a good approximation. Kolakaluri (2013) recommends formulating a distribution function correlating the collection efficiency with particle sizes and subsequently integrating over the efficiencies exhibited by each class. The numerical performance of this is low, due to being calculated in hundreds of thousands of grid cells in every coupling step.

As the calculations may impair performance, this specific implementation first creates an occupancy list of particles in each cell. The same is done for droplet parcels, but automatic pruning is performed for cells where no particles are present. The calculations outlined above are only performed for cells where the droplet parcel occupancy list contains entries.

5.5. Governing Equations in the Eulerian Frame of Reference

In addition to solving the gas kinematics using the Navier-Stokes equations, both vapor/species and heat transport have to be solved in the Eulerian reference frame.

5.5.1. Vapor Transport

Vapor transport is realized by solving the scalar transport equation under the assumption of the Schmidt number Sc = 1:

$$\frac{\partial(\rho_g \alpha_g y_i)}{\partial t} + \nabla \cdot (\phi y_i) = \Delta(\alpha_g(\eta_g + \eta_{g,t}) y_i) + S_{y_i}, \tag{5.23}$$

where y_i represents the vapor mass fraction in the gas phase, ϕ the superficial mass flux of the gas phase, $\eta_{g,t}$ is the turbulent viscosity and S_{y_i} is the sum of all source terms for y_i. The species source term is composed of the contribution by the spray $S_{y_i}^{\text{spray}}$, combustion $S_{y_i}^{\text{comb}}$ and surface liquid evaporation $S_{y_i}^{\text{evap}}$:

$$S_{y_i} = S_{y_i}^{\text{spray}} + S_{y_i}^{\text{comb}} + S_{y_i}^{\text{evap}}, \tag{5.24}$$

while the combustion source term is entirely explicit, the evaporation and spray parts include both explicit and implicit contributions. The remaining inert species y_{inert} is solved for using the expression

$$y_{\text{inert}} = 1 - \sum_{j \neq \text{inert}} y_j. \tag{5.25}$$

Treating the inert specie in any other way, i.e. solving a transport equation as well, would require normalization in the manner of

$$a = \sum_j y_j, \tag{5.26}$$

$$y_i = \frac{y_i}{a}. \tag{5.27}$$

This, in turn, might violate mass conservation of individual species because this information is not reflected in the density/continuity equation. To conserve mass, a source term $\dot{S}_{\rho_g}$ is introduced into the density equation

$$\frac{\partial(\alpha_g \rho_g)}{\partial t} + \nabla \cdot \phi = \dot{S}_{\rho_g}. \tag{5.28}$$

This source term is composed of components for the spray mass source $S_{\rho_g}^{\text{spray}}$ and the evaporation mass source $S_{\rho_g}^{\text{evap}}$:

$$S_{\rho_g} = S_{\rho_g}^{\text{spray}} + S_{\rho_g}^{\text{evap}} \tag{5.29}$$

These terms are treated entirely explicitly due to the segregated treatment of systems of equations in OpenFOAM.

5.5.2. Evaporation Modeling

The Eulerian vapor species source term

$$S_{y_i}^{\text{evap}} = \frac{\sum_j \beta_{j,\text{s}\to\text{f}} A_{\text{s},j} \rho_\text{g}(y_i^* - y_i)}{V} = S_{y_i}^{\text{evap,ex}} + S_{y_i}^{\text{evap,im}} \tag{5.30}$$

for evaporation of surface liquid split between an implicit term, that depends on the current vapor mass fraction and contributes to the diagonal of the linear system

$$S_{y_i}^{\text{evap,im}} = \frac{\sum_j \beta_{j,\text{s}\to\text{f}} A_{\text{s},j} \rho_\text{g}(-y_i)}{V} \tag{5.31}$$

and an explicit component that does instead appear on the right side:

$$S_{y_i}^{\text{evap,ex}} = \frac{\sum_j \beta_{j,\text{s}\to\text{f}} A_{\text{s},j} \rho_\text{g} y_i^*}{V} \tag{5.32}$$

The consequence of this formulation is that the saturation vapor concentration is never exceeded as this would cause the source term to vanish. After solving these, the explicit Lagrangian mass sink $\dot{M}_{\text{g}\to\text{p,evap}}$ is calculated.

5.5.3. Heat Transfer and Transport

Heat transport is described using the energy transport equation

$$\frac{\partial \rho_\text{g} \alpha_\text{g} e}{\partial t} + \nabla \cdot (\phi e) + \frac{\partial \rho_\text{g} \alpha_\text{g} K}{\partial t} + \nabla \cdot (\phi K) + \gamma = \Delta(k_{\text{gp}} e) + S_e, \tag{5.33}$$

where e is either the enthalpy or internal energy, $K = \rho_\text{g}\alpha_\text{g}||\mathbf{U}||^2/2$ is the kinetic energy of the gas phase, γ is the compression work if the system is open ($\gamma = -\frac{\text{d}p}{\text{d}t}$, e is the enthalpy) or $\gamma = \left|\left|\frac{\phi}{\rho_\text{g}}\alpha_\text{g}\mathbf{U}p\right|\right|$ in closed systems if e is the internal energy. In the developed solver γ was chosen to be 0 for better stability. The thermal conductivity k_{gp} is not the pure fluid thermal conductivity but rather that of the particle-fluid suspension as proposed by Syamlal and Gidaspow (1985):

$$k_{\mathrm{gp}} = \frac{1-\sqrt{1-\alpha_{\mathrm{g}}}}{\alpha_{\mathrm{g}}} k_{\mathrm{g}} \tag{5.34}$$

This term includes the improvement of mixing due to particle collision and swarming. The energy source term, S_e, is composed of the contribution by particle heat transfer $S_{e,\mathrm{p}\rightarrow\mathrm{f}}$, by the spray $S_{e,\mathrm{spray}\rightarrow\mathrm{f}}$ and the surface liquid film $S_{\mathrm{e,s}\rightarrow\mathrm{f}}$:

$$S_e = S_{e,\mathrm{p}\rightarrow\mathrm{g}} + S_{\mathrm{e,s}\rightarrow\mathrm{g}} + S_{e,\mathrm{spray}\rightarrow\mathrm{g}} \tag{5.35}$$

At first glance, one may miss an energy source term for evaporation. This is due to the implementation of thermodynamics in OpenFOAM: a mass source, like vapor entering the gas phase, will change both the composition and density - and thus increase the heat capacity without changing the temperature.

Neglecting the spray contribution, implicit-explicit splitting can also be performed for S_e:

$$S_{\mathrm{e}} = \frac{\sum_j \alpha_{j,\mathrm{p}\rightarrow\mathrm{f}} A_j \rho_{\mathrm{g}} (T_{j,\mathrm{sp}} - T_{\mathrm{g}})}{V} = S_{\mathrm{e}}^{\mathrm{ex}} + S_{\mathrm{e}}^{\mathrm{im}}, \tag{5.36}$$

with the mean particle surface temperature $T_{j,\mathrm{sp}} = \varphi_{\mathrm{wet},j} T_{j,\mathrm{s}} + (1-\varphi_{\mathrm{wet},j}) T_{j,\mathrm{p}}$.

The explicit part of this source term is the fixed part containing the particle temperature

$$S_{\mathrm{e}}^{\mathrm{ex}} = \frac{\sum_j \alpha_{j,\mathrm{p}\rightarrow\mathrm{g}} A_j \rho_{\mathrm{g}} T_{j,\mathrm{sp}}}{V} \tag{5.37}$$

and the implicit part is the gas temperature-dependent part

$$S_{\mathrm{e}}^{\mathrm{im}} = \frac{\sum_j \alpha_{j,\mathrm{p}\rightarrow\mathrm{g}} A_j \rho_{\mathrm{g}} (-T_{j,\mathrm{g}})}{V}. \tag{5.38}$$

For use in the energy equation, these sources have to be converted to enthalpy/internal energy by multiplying with the heat capacity and correcting for the thermodynamic reference temperature.

As with the vapor phase, the explicit Lagrangian source terms are calculated after solving the Eulerian flow, implying conservation of mass among both phases.

A verification of the implementation can be found in appendix A.

5.6. Solver Validation against Drying Experiments

For validation, the Glatt GF3 plant used in conjunction with 600 µm silica glass particles was chosen. A schematic of this plant is shown in Figure 5.5.

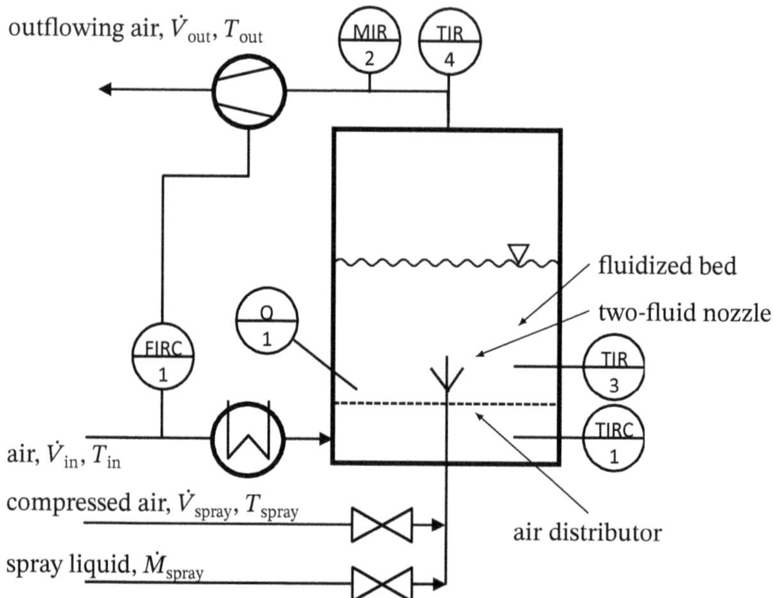

Fig. 5.5.: Schematic of the Glatt GF3 apparatus.

The plant was operated with a *Schlick Mod. 970/S4* two-fluid nozzle at an operating pressure of 2 bar, corresponding to an atomization air flow rate of $5\,m^3\,h^{-1}$ at the cap setting 2. All specified fluidization air volumetric flow rates are specified at standard conditions, and prescribed using the corresponding mass flow rates in the simulations. A variety of different process conditions and transient changes were applied to ensure correct reproduction of flow regime changes as well as system latencies.

The material properties and contact model parameters were taken from the material property database of the MUSEN DEM code (Dosta and Skorych, 2020) and can be found in Tab. 5.2, together with the numerical setup. The validation parameter of choice is the particle temperature, as particle surface liquid mass measurement has proven difficult due to the high cohesivity of wetted particles. This makes sample extraction more difficult. Non-porous particles also have very low water loadings, reaching the limits of the available gravimetric measurement technique. Outlet temperatures and humidities just above the particle bed are determined by integral balances and thus of no benefit.

Tab. 5.2.: Numerical parameters, material properties and contact model coefficients used on the DEM side of the validation case.

Numerics		
Time Step		
CFD	Δt_{CFD}	$5 \cdot 10^{-4}\,\mathrm{s}$
DEM	Δt_{DEM}	$5 \cdot 10^{-5}\,\mathrm{s}$
Scaling Factor (Coarse Graining)	δ_{CG}	2
Particle		
Diameter	d_{p}	$1 \cdot 10^{-3}\,\mathrm{m}$
Density	ρ_{p}	$2500\,\mathrm{kg\,m^{-3}}$
Young's Modulus	Y_{p}	1 MPa
	Y_{w}	1 MPa
Poisson Ratio	ν_{p}	0.22
	ν_{w}	0.22
Coefficient of Restitution	$e_{\mathrm{p\text{-}p}}$	0.75
	$e_{\mathrm{p\text{-}w}}$	0.62
Coefficient of Friction	$\eta_{\mathrm{fr,p\text{-}p}}$	0.3
	$\eta_{\mathrm{fr,p\text{-}w}}$	0.3
Coefficient of Rolling Friction	$\eta_{\mathrm{rfr,p\text{-}p}}$	0.025
	$\eta_{\mathrm{rfr,p\text{-}w}}$	0.025
Heat Capacity	$C_{\mathrm{v,p}}$	$840\,\mathrm{J\,kg^{-1}\,K^{-1}}$
Liquid		
Density	ρ_{s}	$1000\,\mathrm{kg\,m^{-3}}$
Heat Capacity	$C_{\mathrm{v,s}}$	$4186\,\mathrm{J\,kg^{-1}\,K^{-1}}$
Heat of Evaporation	$\Delta h_{\mathrm{s}}^{\mathrm{LV}}$	$2.5 \cdot 10^{6}\,\mathrm{J\,kg^{-1}}$
Diameter	d_{droplet}	$20\,\mu\mathrm{m}$
Injection Rate	$\dot{N}_{\mathrm{droplet}}$	$1 \cdot 10^{5}\,\mathrm{s^{-1}}$
Gas		
Density		Ideal Gas Law (appendix C.1)
Molecular Weight	$M_{\mathrm{w,N_2}}$	$28.0134\,\mathrm{g\,mol^{-1}}$
	$M_{\mathrm{w,H_2O}}$	$18.015\,\mathrm{g\,mol^{-1}}$
Viscosity		Sutherland (1893) Law (appendix C.3)
Parameter A_{s}	$A_{\mathrm{s,N_2}}$	$1.672\,12 \cdot 10^{-6}\,\mathrm{J\,kg^{-1}}$
	$A_{\mathrm{s,H_2O}}$	$1.672\,12 \cdot 10^{-6}\,\mathrm{J\,kg^{-1}}$
Parameter T_{s}	$T_{\mathrm{s,N_2}}$	170.672 K
	$T_{\mathrm{s,H_2O}}$	170.672 K
Coupling		
Drag Law		Beetstra et al. (2007)
Coupling Interval	$\Delta t_{\mathrm{coupling}}$	$5 \cdot 10^{-4}\,\mathrm{s}$

5.6.1. Case Analysis

This case exhibits exemplary fluidized bed behavior with respect to particle heating. In Fig. 5.6, the temperature of the air entering and leaving the granulator and the mean particle temperature are plotted over time.

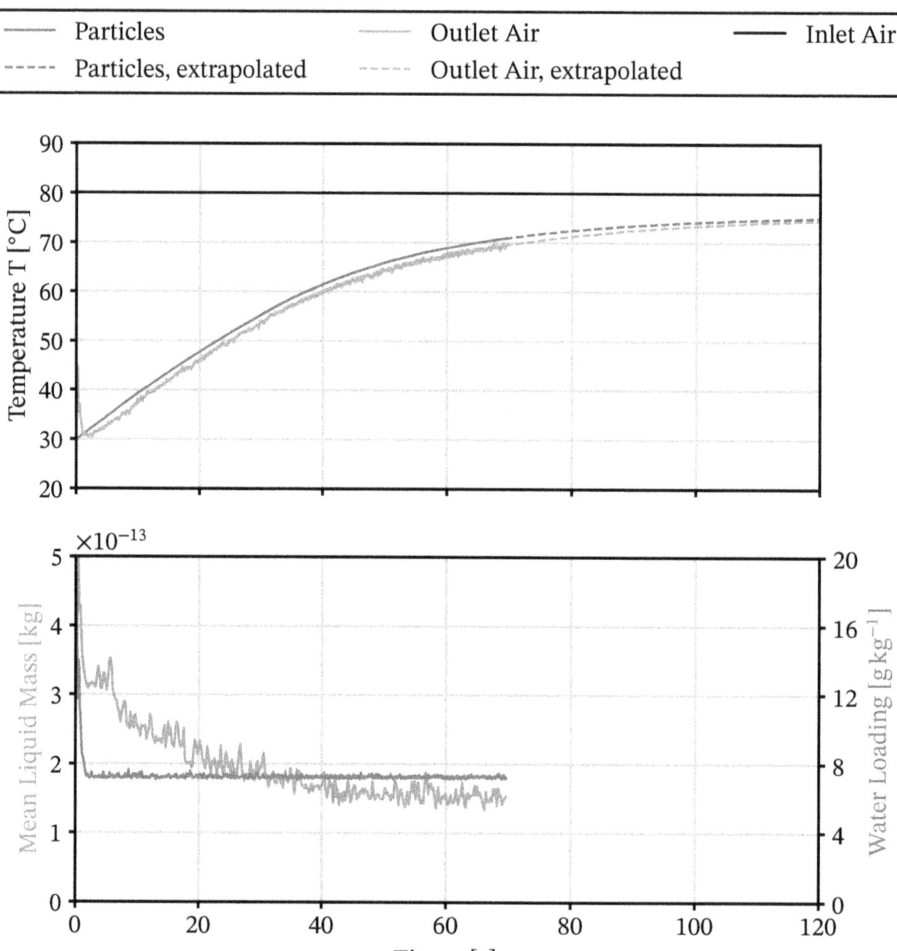

Fig. 5.6.: Particle and outlet air temperatures for an example simulation case. The equation for the extrapolated outlet air temperature is $T_{\mathrm{air}}(t) = -56.8\,°\mathrm{C} \cdot \exp(-t/31.5\,\mathrm{s}) + 75.8\,°\mathrm{C}$ and the equation for the particle temperature is $T_{\mathrm{p}}(t) = -62.4\,°\mathrm{C} \cdot \exp(-t/27.2\,\mathrm{s}) + 75.8\,°\mathrm{C}$.

The particle temperature closely follows the temperature of air leaving the system, as can be expected from the model of plug flow of air through an ideally mixed system of particles that is commonly used to describe fluidized beds. The temperatures can be described by the solution of a first-order differential equation $T(t) = a \cdot \exp(-t/\tau) + T_\infty$, where a is the initial temperature difference. The asymptotic temperature T_∞ signifies the temperature that will be assumed if the experiment or simulation is run for an indefinite time. The asymptotic temperature of the outlet air has a constant difference to the inlet air, indicating that the evaporation decreases the air temperature as it passes through the system.

The time constant τ is the duration it takes for a system to overcome 63.2% of the temperature difference a between the initial temperature, given by $T(0) = a + T_\infty$, and the asymptotic temperature T_∞. Given the high heat transfer coefficients in fluidized beds, the time constant of the mean particle temperature is only slightly lower than that of the outflowing air. This is plausible, since heat transfer requires a temperature gradient to be present.

The corresponding water vapor fraction in the outflowing air drops to a very low value of about $0.5\,\mathrm{g\,kg^{-1}}$. This is much lower than the saturation vapor fraction of $291\,\mathrm{g\,kg^{-1}}$ at the asymptotic temperature, showing this operating point to be dominated by convection rather than saturation. The water mass on particles' surfaces in the system shows similar behavior as the particle and outlet temperatures, albeit with a time constant of 14.8 s. As the particles were pre-wetted with a total water mass of $4.77 \cdot 10^{-4}$ kg of total liquid, the drop instead of rise of the water content can be explained.

5.6.2. Validation against Steady-State Experiments

To establish the ability of this implementation to accurately describe evaporation, the equilibrium particle temperatures of systems with continuous liquid injection are recorded in a set of laboratory experiments. An overview of the experiments and their results are given in Tab. 5.3. Volume flow rates, temperatures of inflowing air and water spray rates were systematically varied, and one experiment was run three times to quantify the reproducibility of the experiments. A parity plot of experimental and simulation particle temperatures, both determined by fitting $T_\mathrm{P}(t) = a \cdot \exp(-t/\tau) + T_{\mathrm{P},\infty}$, is displayed in Fig. 5.7. Here, a good agreement at a maximum deviation of the simulation-derived mean particle temperatures of less than 10% from the experimentally measured bed temperatures can be observed. All of the simulation temperatures lie higher than the measured ones. This indicates a systematic error rather than a random one.

Tab. 5.3.: Process conditions and resulting particle temperatures under steady-state conditions. Note that these experiments were conducted with 5 $Nm^3 h^{-1}$ atomization air.

Process Conditions			Particle Temperature $T_{P,\infty}$ [°C]	
$\dot{V}_{in}$ [$Nm^3 h^{-1}$]	T_{in} [°C]	$\dot{M}_{spray}$ [$g\,min^{-1}$]	Simulation	Experiment
120	70	0.0	69.2	64.5
160	60	18.1	46.0	43.7
180	60	18.1	47.3	44.6
200	50	18.2	38.5	36.1
200	60	18.1	48.5	45.3
200	60	19.0	48.0	46.0
200	70	0.0	69.5	65.2
200	70	0.0	69.5	67.4
200	70	0.0	69.5	64.6
200	70	18.7	57.9	54.4
200	70	43.8	42.0	41.2
200	80	18.9	67.6	63.7
200	90	0.0	89.3	81.8
200	90	18.8	77.6	73.0

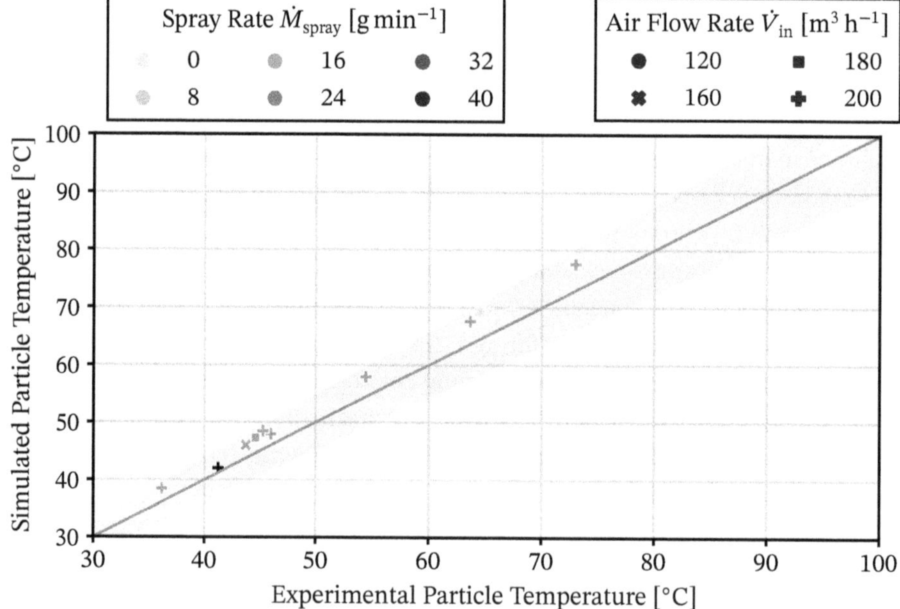

Fig. 5.7.: Parity plot of particle temperatures in both simulation and experiments for the steady-state case. The striped region is the bound of 10% deviation.

Figure shows the relationship between the particle temperature and fluidization air temperature, spray rate and fluidization air temperature. The deviation of temperatures between experiments and simulations decrease with both higher spray rates and lower air temperatures. This suggests that the assumption of adiabatic walls in simulations introduces a low error. Another reason can be that the measurement was performed by a Pt-100 element that is directly introduced into the bed. Performing the measurements like this does not directly yield the particle temperature but a mixture of the gas temperature and particle temperature.

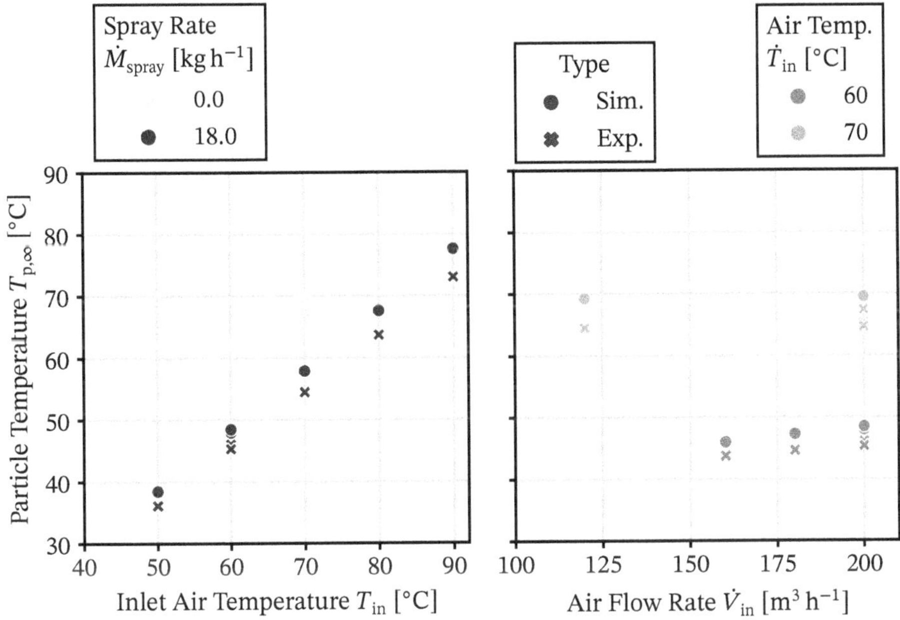

Fig. 5.8.: Particle temperatures in steady-state simulation and experiment under variation of the fluidization air temperature and the fluidization air flow rate. Note that the experiment with no spray (0 g min^{-1}) was conducted three times to assess the variability of the system.

5.6.3. Validation of Mean Temperature Response against Transient Experiments

Three different scenarios were considered: Two spray rate changes at two different bed masses and fluidization air flow rates and one where the fluidization air temperature is changed, as outlined together with the experimental and simulated time constants in Tab 5.4.

Tab. 5.4.: Process conditions, their changes and the resulting time-constant of the change in particle temperature.

Process Conditions				Time Constant	
Bed Mass M[kg]	Air Flow Rate $\dot{V}_{in}$[m^3 h^{-1}]	Temperature T_{in}[°C]	Spray Rate $\dot{M}_{spray}$ [g min^{-1}]	Sim $\tau(T_P)$ [s]	Exp $\tau(T_P)$[s]
5	120	70	$0 \rightarrow 21.6$	147	143
2.5	200	70	$0 \rightarrow 43.4$	34	45
2.5	200	$80 \rightarrow 70$	34	32	33

The experimental time constant at the operating point of 120 m^3 h^{-1} was subject to the response time of the Pt-100 thermoelement, as, at this volume flow rate, the system is barely fluidized and the heat transfer coefficient is low. To make the simulation-derived time constants comparable, the signal response curve (see Fig. 5.9) was determined by inserting a cool thermoelement into a running system. By performing a convolution with that response curve before determining the time constant, the effect of added measurement latency can be compensated.

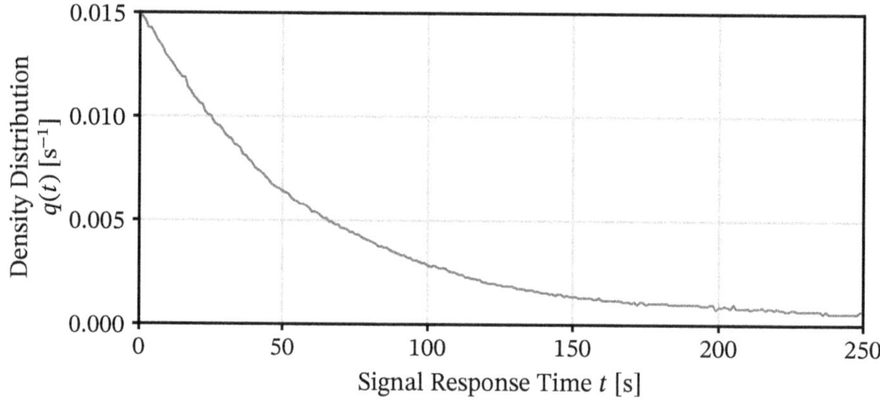

Fig. 5.9.: Normalized response time of a Pt-100 thermoelement when being inserted into a fluidized bed just above the point of minimum fluidization.

The agreement with regard to the time constant is perfect for the two operating points with a low spray rate, deviating less than 4%. When instantaneously switching to a high spray rate of 43.4 g min^{-1}, the time constant of experiments is 32% higher, indicating a substantial change in hydrodynamics that reduces mixing.

5.7. Summary

Two types of surface coverage modeling techniques, one probabilistic-discretized and another one analytical, were evaluated. The results of the probabilistic-discretized approach was shown to be representable by the analytical ones. The effect of hysteresis in evaporation could only be captured in surface discretized coverage.

The mathematical foundations of the interaction of particle, surface liquid, spray and gas phase in drying were formulated and the numerical intricacies elaborated upon. A solver capable of capturing the basic physics of fluidized bed spray granulation and drying was developed and verified for mass and energy conservation.

A thorough validation of this solver against experiments was performed. The solver showed agreement with respect to both system latency in transient experiments and particle temperature in steady-state experiments.

6 Development of a Novel Approach for Prediction of Particle Properties in Fluidized Bed Spray Granulation

The goal of this chapter is the development of a method to predict the surface structures of particles produced by fluidized bed spray layering granulation using the CFD-DEM method. To this end, a simple state-variable/event tracking approach was developed. The state of the droplet at the time of impact on the particle surface, as well as the time required for drying, is correlated to product properties that quantify surface structure morphology such as porosity. This approach was tested on a demonstration case from literature, where a particle core is coated with sodium benzoate solution.

This chapter is structured in the following manner: First, a brief overview of the physics involved in fluidized bed spray granulation is given. The developed tracking approach is presented with the physical justification based on the theoretical overview. Then, the approach is demonstrated based on the aforementioned demonstration case.

6.1. Theoretical Considerations

The properties of products of fluidized bed spray granulation are highly dependent on the corresponding drying conditions, as well as on the way in which the spray is injected.

In this section, previous macroscopic experimental studies will be reviewed, and an overview of the supposed underlying physics will be given.

6.1.1. Macroscopic Experimental Studies

The most prominent experimental studies in this area were performed by Hoffmann (2016), Hoffmann et al. (2015), Rieck et al. (2015) and Schmidt et al. (2017a).

Rieck et al. (2015) and Hoffmann (2016) studied the deposition of soluble salts (sodium benzoate) on particle cores that are made of either silica glass or alumina (γ-Al_2O_3).

They varied the solution spray rate and fluidization air temperature in batch granulation experiments and analyzed the porosity of the deposited salt layer using micro-computer tomography (μ-CT). A linear correlation between the drying potential of the inflowing fluidization air and the resulting layer porosity was identified.

Schmidt et al. (2017a) performed similar experiments using a suspension of solid fines as a spraying liquid and varied (1) the atomization pressure, (2) the fluidization air temperature and (3) the spray rate.

Diez et al. (2018) performed continuous granulation experiments in a pilot-scale fluidized bed spray granulator equipped with a mill-sieve cycle. This allowed for the production of granules whose entire structure was generated using the process conditions in the granulator. They analyzed

- the product moisture content,
- the area surface roughness as analyzed by confocal microscopy,
- the modulus of elasticity using compression testing,
- the granule porosity using X-ray micro computer tomography,
- the wetting behavior using the contact angle as given by the sessile drop experiment.

The operating point with respect to gas mass flow rate was kept constant while varying spray rate and air temperature to capture the effect of the drying potential. The overall results fit with the findings of Schmidt and Hoffmann: Granule porosity increases for *wetter* drying conditions, equating to higher spray rates, lower drying potentials and lower temperatures. As this study considered actual target properties like compression strength and wetting behavior as well, it is of great interest for this work. All of the relationships between these and the porosity behave as expected - compression strength increases with *harsher* drying conditions (= high drying potential), wetting behavior improves with *wetter* drying conditions due to increasing surface roughness/porosity.

Other studies focus on agglomeration of primary particles into agglomerates - the physics involved there do not apply for layering granulation/coating and are thus not considered here.

6.1.2. Micro-Scale Processes in Layering Granulation

The underlying physics of spray layering granulation can be divided into three distinct phase, as illustrated in Fig 6.1:

1. microprocesses in droplets,
2. impingement of droplets on the particle surface,
3. microprocesses on the particle surface.

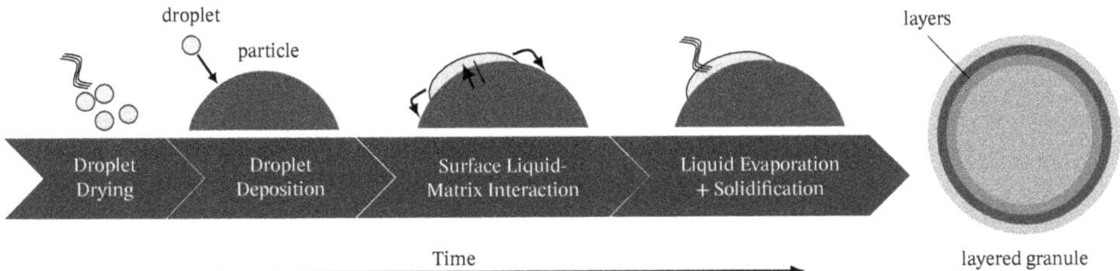

Fig. 6.1.: Illustration of the succession of microprocesses that occur in fluidized bed spray granulation when taking place in the layering regime.

As mentioned by Rieck et al. (2015) and Hoffmann (2016); Hoffmann et al. (2015), the nucleation processes occurring in a droplet can be assumed to have a determining influence on the surface structure. These processes are studied in depth for spray drying (de Souza Lima et al., 2020; Zang et al., 2019), where the droplets are dried to completion and the product is chiefly determined by the removal of solvent by evaporation. Mechanistically, the structures created by suspensions and solutions are distinguished: Suspensions are dominated by the competition of solvent removal, capillary forces as well as interlocking/short-range forces of the constituents. Solutions have the added complexity of nucleation and crystal growth.

The timescale on which solvent removal takes place determines the process. For solutions, fast solvent removal prohibits the slow growth of larger crystals after formation and, ultimately, results in smoother layers on the particles. For suspensions, the interplay of phenomena is more complex: Schmidt et al. (2017b) judged the rate of solvent removal to be less important than the final droplet solids concentration - altering the droplet's viscosity and size. These later lead to differences in droplet spreading on the surfaces.

Upon impact on the particle surface, the droplet will either stick without rupture, splash because of the high relative velocities, or rebound back into the gas phase. One model describing this is the one by Bai and Gosman (1996) that uses the normal Weber number

$\mathrm{We_n} = \frac{\rho_{\mathrm{drop}} U_{\mathrm{n},0}^2 d_{\mathrm{drop}}}{\sigma_{\mathrm{drop}}}$ and the normal Reynolds number $\mathrm{Re_n} = \frac{\rho_{\mathrm{drop}} v_{\mathrm{n},0} d_{\mathrm{drop}}}{\mu_{\mathrm{drop}}}$ to differentiate between the following impact regimes:

$$\begin{cases} \text{stick} & \mathrm{We_n} \leq 5 \\ \text{rebound} & 5 < \mathrm{We_n} \leq 10 \\ \text{spread} & \mathrm{We_n} > 10 \wedge \mathrm{We_n}\mathrm{Re_n^{0.5}} < H_{\mathrm{cr}} \\ \text{splash} & \mathrm{We_n}\mathrm{Re_n^{0.5}} \geq H_{\mathrm{cr}} \end{cases} \tag{6.1}$$

With the sticking criterion

$$H_{\mathrm{cr}} = \left(1500 + 650\left(\frac{h_{\mathrm{roughness}}}{d_{\mathrm{droplet}}}\right)^{-0.42}\right)\left(1 + 0.1\mathrm{Re_n^{0.5}}\min\left(\frac{h_{\mathrm{film}}}{d_{\mathrm{droplet}}}, 0.5\right)\right) \tag{6.2}$$

according to Han et al. (2000) depending on the Reynolds number $\mathrm{Re_n}$, the surface roughness $h_{\mathrm{roughness}}$ and the height of the existing film h_{film}.

This step is of importance for the successive behavior of the liquid layer on the surface, as it determines the contact area with both gas and particle surface - thus determining the starting point for micro-scale processes acting on the liquid and particle surface.

On the particle surface, there are four main ways in which the resulting liquid layer may evolve over time:

- evaporate to the surrounding bulk gas,
- spread on the surface,
- get imbibed by the particle matrix,
- dissolve the particle matrix.

As for evaporation, this process can be accurately described using heat/mass transfer correlations, assuming that the relevant area is accurately modelled (Gunn, 1978):

$$\dot{M}_{\mathrm{film}\rightarrow\mathrm{gas}}^{\mathrm{evap}} = \beta_{\mathrm{film}\rightarrow\mathrm{gas}} A_{\mathrm{film}} \rho_{\mathrm{gas}} (y_{\mathrm{film}}^* - y_{\mathrm{gas}}) \tag{6.3}$$

This is elaborated upon in section 5.2.

The spreading behavior of a droplet can be approximated using a variety of different correlations. Ogarev et al. (1974) derived the following relationship for the time-dependent radius of the liquid film:

$$R(t) = \frac{24V_{\text{film}}\sigma(\cos(\theta_0) - \cos(\theta_\infty))}{\pi\eta}t^{1/4} \tag{6.4}$$

Chua et al. (2011) gives an approximation of the model of Clarke et al. (2002) that was derived for a droplet on a flat surface:

$$\frac{dR}{dt} = \frac{d_0^3\sigma\cos(\theta(R))}{3\eta}R^{-3} \tag{6.5}$$

Here, a closure for the relationship between the contact angle $\theta(R)$ at any given radius R is required.

The dissolution of the particle matrix by the liquid film is deemed another important phenomenon, especially when salt solutions of concentration far below the solubility limit are concerned. Noyes and Whitney (1897) investigated this phenomenon and found an expression for the dissolution rate

$$\dot{M}_{\text{diss,p}\rightarrow\text{film}} = \frac{D_{\text{l,s}}A_{\text{film}}}{h_{\text{film}}}(\rho_{\text{s,film}}x^*_{\text{s,film}} - \rho_{\text{s,film}}x_{\text{s,film}}) \tag{6.6}$$

dependent on the diffusion coefficient $D_{\text{l,s}}$ and the concentration at the solid-liquid interface $\rho_{\text{s,film}}x^*_{\text{s,film}}$. Spreading and dissolution processes usually coincide with imbibition of the surface film into the particle layer.

6.2. Proposed Approach

In this work, an approach to estimate the properties of products of fluidized bed spray granulation based on tracking simple quantities in CFD-DEM simulations is proposed.

6.2.1. Requirements

The developed approach aims to satisfy the following requirements:

- *predictiveness*: The approach should make accurate predictions for scaled-up systems. This requires that the set of tracked quantities should be sufficient to account for the involved microprocesses that determine product properties.
- *robustness and versatility*: The approach should track quantities that normally influence product properties and be applicable to a wide variety of target systems. A calibration workflow should be able to be adapted to different product properties.

- *performance*: The approach should not imbue a large computational overhead or require the reduction of DEM timesteps to resolve microprocesses taking place on a per-particle basis - this would preclude the application to cases that have long timescales. The approach may not require large amounts of per-particle state information that makes the treatment of industrially relevant systems impossible.

6.2.2. Tracked Quantities

The tracked quantities are of two distinct types, those that are *per-particle* and those that are *per-event*.

Per-particle quantities are scalar values or vectors associated with an individual particle, for example a residence time in a specific region, or the temperature. *Per-event* quantities are calculated as they occur, like the contact force in the case of particle collisions. Tracking many different per-particle properties will substantially slow down a simulation. They have to be kept in memory during the entire duration of the simulation and have to be written to disk for visualization or to create a checkpoint. This either requires stalling the simulation or spawning a thread that accesses the disk. Because the quantity is present for every particle, this creates a great demand for disk storage space and working memory in the computer used - one double precision floating point number for every particle in the system in the case of scalars. Even for particles that are, for example, not wetted, this information is written to disk and kept in memory while not immediately contributing to the statistics. Simulation performance will decrease as these quantities have to be communicated when crossing from one processors simulation domain to another one, creating parallelization overhead. Because they incur both a performance penalty and demand a lot of storage space per-particle properties should be chosen sparingly with care.

In contrast, an almost arbitrary number of *per-event* quantities can be recorded. For these, data can be written to disk for every relevant event as it occurs, specific to the quantity and to the event that is tracked. For this reason, *per-event* quantities should be preferred over *per-particle* quantities.

For the category of per-particle quantities, tracking the lifetime of the liquid film on the particle surface is an obvious choice in fluidized bed spray granulation. The duration a liquid film is present should, in theory, be able to capture the interplay of evaporation, imbibition/dissolution, and spreading. It also correlates to the kinetics of crystallization of soluble components of the liquid film.

As for per-event properties, the following properties should be tracked:

- *solids concentration in the droplets at impact*: The concentration of the solid component of droplets upon impact is an indicator for the intensity of the drying conditions that occur - determining interplay of solvent removal and aggregation of the solid components by diffusion, relating to the time available for nucleation and crystallization to occur for solutions and aggregation to take place in the case of suspensions.
- *relative velocity at impact*: The relative velocity between particle and droplet, together with viscosity and surface tension (both dependent on solids concentration) should correlate with the droplet interaction regime, in accordance with the impingement model outlined in section 6.1.2.

6.2.3. Evaluation of the Simulations

The subsequent evaluation has to be considered as well. Two choices may be made:

- *population-based*: Distributions over all tracked quantities are analyzed separately. This has the key advantage of being easily automated and taking into consideration the spread of the entire population of particles.
- *particle-based*: For single or selected particle populations, tracked quantities are correlated with each other to give a temporal sequence of events or states in which the particle is, e.g. periods of drying alternating with wetting. While this may be more intuitive to analyze, automatically scaling this analysis to the entire particle population is much more difficult.

For these reasons the first approach is more desirable. The simulations are evaluated on population bases.

6.2.4. Workflow

In this section, a workflow is proposed that takes the following form:

1. calibration experiments
2. calibration simulations
3. evaluation of simulations and experiments, derivation of a mapping
4. predictive simulation

It is given in schematic form in Fig. 6.2.

The first step must always lie in a set of laboratory scale granulation experiments that create conditions. Ideally, a wide variety of product properties are generated. Otherwise, more practically, an acceptable range of products are generated.

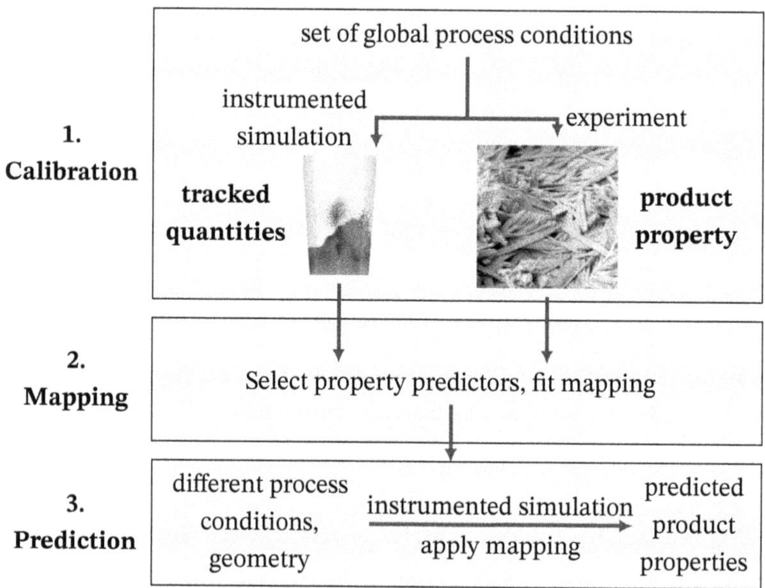

Fig. 6.2.: Schematic of the proposed approach to correlate the tracked-quantity with product-property.

Extrema of the process conditions must be found that still result in an acceptable product, with a few experiments set at intermittent conditions to ensure sufficient understanding of the micro-scale phenomena. The experiments must then be replicated in simulations while tracking the set of quantities that are suspected to be descriptive of processes that determine the product properties. For matching process conditions, the resulting macroscopic, experimentally determined product properties and tracked quantities have to be correlated so that a mapping can be derived. Ideally, this step gives insight into what physics play a role on the microscopic scale by exposing the sensitivity of certain product properties to the physical meaning of the tracked quantities. Finally, the application of the model mapping can serve either the purpose of confirming that a new design of a bigger apparatus creates conditions that will result in similar product properties or to gain insight what changes in geometry or operating conditions will have an influence on product properties.

6.2.5. Limitations and Assumptions

By relying on macroscopic calibration experiments, the resulting mapping will inevitably be limited to the range of conditions that droplets and particles experience - these will not necessarily match the conditions present in the apparatus that is to be designed. Here, a distinction must be made between the macroscopic conditions and those experienced by the droplets - the latter are relevant to the developed approach. The only remedy will be that upon analysis of the distributions of the tracked quantities these will appear to be outside of the range of conditions that were covered by the calibration, showing the need for further calibration experiments, simulation and a new mapping.

More crucially, the wrong choice in tracked quantities will result in wrong predictions, as the mapping between tracked quantities and predictions relies on the causal dependency of tracked quantities in the micro-processes that give rise to the product properties.

6.3. Derivation of a Mapping between Tracked Quantities and Product Properties

The method for deriving a mapping between tracked quantities and product properties should fulfill the following set of criteria:

- *general*: This method should be applicable to a wide range of physical phenomena and therefore be formulated in a manner that applies to a wide range of situations.
- *robust*: Parameter selection, at this stage, should not require fine-tuning or manual intervention. The data itself should be the only foundation for selecting parameters.
- *meaningful*: The mapping itself should give insight into the process at hand, by revealing the sensitivities of product formation to microscopic events.

6.3.1. Fitting of a Linear Regression Model

Mathematically speaking, the set of tracked quantities resulting from a simulation for an operating point a can be described as $\{\mathbf{X}^m, \mathbf{X}^n, ...\}_a$ where $\mathbf{X}^m = \{X_i^m\}$ is an individual per-event value, the subscripts $i, j, ...$ are indices counting over each respective tracked event and the superscripts $m, n, ...$ refer to the kind of events that occurs. The resulting experimentally obtained product property from the corresponding process conditions in a is referred to as y_a.

A mapping $f(\{\mathbf{X}^m, \mathbf{X}^n, ...\}_a) \rightarrow y_a$ is therefore to be derived. The distribution of any of the tracked quantities $\{\{X_i^m\}, \{X_j^n\}, ...\}_a$ for each simulation has to be reduced to a set of descriptive statistical quantifiers. This can be either a set of percentiles or statistical moments, most prominently of which are the mean $\mu_0(\mathbf{X})$, the variance $\mu_1(\mathbf{X})$ and the skewness $\mu_2(\mathbf{X})$:

$$\mu_0(\mathbf{X}^m) = \frac{1}{N_i} \sum_{\text{events } i}^{N_i} (X_i^m) \tag{6.7}$$

$$\mu_1(\mathbf{X}^m) = \frac{1}{N_i} \sum_{\text{events } i}^{N_i} (X_i^m - \mu_0(\mathbf{X}^m))^2 \tag{6.8}$$

$$\mu_2(\mathbf{X}^m) = \frac{1}{N_i} \sum_{\text{events } i}^{N_i} (X_i^m - \mu_0(\mathbf{X}^m))^3 \tag{6.9}$$

Where N_i is the number of events of type m that were tracked. The result is a set $\mathbf{M}_a$ of scalar statistical quantifiers for each simulation:

$$\mathbf{M}_a = \{\mu_0(\mathbf{X}^m), \mu_1(\mathbf{X}^m), \mu_2(\mathbf{X}^m), \mu_0(\mathbf{X}^n), \mu_1(\mathbf{X}^n), \mu_2(\mathbf{X}^n), ...\} \tag{6.10}$$

After transforming each of the distributions to a set of percentiles and/or statistical moments, a relation $\hat{f}(\mathbf{M}_a) \rightarrow y_a$ can be attempted using the linear ansatz:

$$\hat{f}(\mathbf{M}) = \mathbf{M} \cdot \boldsymbol{\alpha} + y_0 \tag{6.11}$$

Subsequently, the data is normalized by subtracting the mean value of the distribution and dividing by its standard deviation. This becomes important when feature selection using L_1 regularization is introduced. The coefficients $\boldsymbol{\alpha}$ and the intercept y_0 can be determined using the least-squares method that minimizes the objective function

$$g(\boldsymbol{\alpha}, y_0) = \sum_{\text{process conditions } a} (\mathbf{M}_a \cdot \boldsymbol{\alpha} + y_0 - y_a)^2 \tag{6.12}$$

While this approach will result in an optimal fit, it fails to consider the effect of over-fitting that is introduced by using a wide variety of parameters.

The *least absolute shrinkage and selection operator* (Lasso) (Pedregosa et al., 2011) regression model avoids this issue by penalizing under the 1-norm:

$$g(\boldsymbol{\alpha}, y_0) = \sum_{\text{process conditions } a} (\mathbf{M}_a \cdot \boldsymbol{\alpha} + y_0 - y_a)^2 + \lambda ||\boldsymbol{\alpha}||_1 \tag{6.13}$$

The regularization hyper-parameter $\lambda > 0$ ensures that overfitting carries a penalty and yields sparse coefficients, which can be used for selecting only the most important parameters in the mapping. This parameter therefore blends between reducing the number of factors and fulfilling the original objective function (6.12). A value of beta that is too large will result in the artificial depression of coefficients, while a low value will allow for overfitting.

To ensure a balance of both, cross-validation can be performed while fitting. One such approach is k-fold splitting of the dataset, performing a fit on all but one fold and validating using the remaining folds. The value of λ that minimizes the error is the correct choice.

6.4. Demonstration Case according to Hoffmann (2016)

In this section, an application case is explored to evaluate all but the last step of the workflow outlined in section 6.2.4. The tracked-quantity approach is tested for the processes of spray granulation with a salt solution. The used experimental plant is a modified *Glatt GPCG 1.1* that is illustrated in Fig. 6.3, having a cylindrical geometry with a diameter of 150 mm and a *Schlick Mod. 970/0 S4* externally mixing two-fluid spraying nozzle in top-spray configuration at a distance of 200 mm to the distributor plate. The fluidization air is always preconditioned to an absolute humidity of $y_{H_2O,in} = 1\,g\,kg^{-1}$.

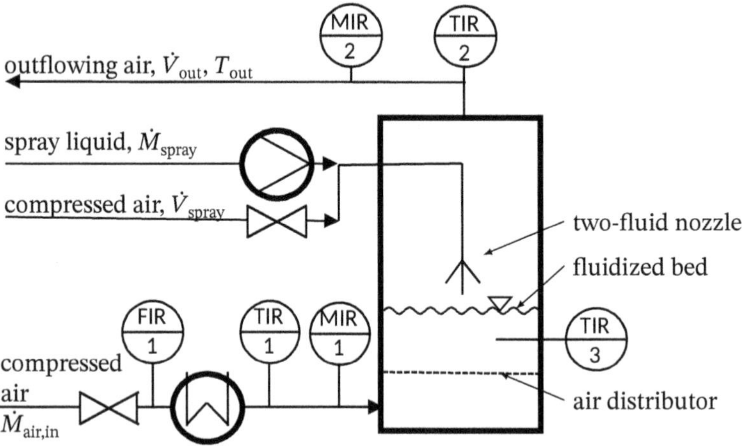

Fig. 6.3.: Schematic of the modified *Glatt GPCG 1.1* apparatus used in the experiment (adapted from Hoffmann (2016)).

This case corresponds to experiments E10-E13 described by Hoffmann (2016) and the process conditions that were varied, as well as the resulting shell porosities, are given in Tab. 6.1. Here, a sodium benzoate solution in water is sprayed onto glass particles. The varied process conditions are the temperature of the inflowing air and the spray rate, whereas solids concentration is kept constant at 30%. The analysis of the resulting particle porosity was done by X-ray computer tomography of single particles.

Tab. 6.1.: Process conditions and experimentally determined shell porosities for the demonstration case (Hoffmann, 2016).

Name	Air Temperature [°C]	Spray Rate [$kg\,h^{-1}$]	Experimental Porosity
E10	50	0.504	0.54
E11	50	0.967	0.64
E12	95	0.512	0.46
E13	95	1.277	0.50

6.4.1. Case Setup

This case is set up in the same manner as the Schmidt et al. (2017b) case in chapter 7. The particle, spray liquid and gas phase properties and numerical setup are shown in Tab. 6.2. The gas phase was assumed to be an ideal gas that consists of Nitrogen gas and water vapor, with the heat capacity calculated using a mixed approach from the NASA fourth-order temperature-dependent polynomials and the viscosity calculated using the Sutherland (1893) equation. Nozzle air is injected at a rate of 1.7 kg h^{-1} and fluidization air at 120 kg h^{-1}. Simulations were run for a simulation time of 150 s.

Tab. 6.2.: Material properties and contact model parameters of the demonstration case (Hoffmann, 2016).

Particle		
Diameter	d_p	$5 \cdot 10^{-4}$ m
Density	ρ_p	2500 kg m^{-3}
Heat Capacity	C_p	840 Jkg^{-1}K^{-1}
Youngs Modulus	Y_p	$1 \cdot 10^6$ Pa
	Y_w	$5 \cdot 10^8$ Pa
Poisson Ratio	η	0.22
Restitution Coefficient	e_{pp}	0.75
	e_{pw}	0.62
Friction Coefficient	$k_{fr,pp}$	0.3
	$k_{fr,pw}$	0.3
Rolling Friction Coefficient	$k_{rfr,pp}$	0.025
	$k_{rfr,pw}$	0.025
Gas		
Density	Ideal Gas Law	
Molecular Weight	M_{w,N_2}	28.0134 g mol^{-1}
	M_{w,H_2O}	18.015 g mol^{-1}
Viscosity	Sutherland (1893) Law	
Parameter A_s	A_{s,N_2}	$1.672\,12 \cdot 10^{-6}$ J kg^{-1}
	A_{s,H_2O}	$1.672\,12 \cdot 10^{-6}$ J kg^{-1}
Parameter T_s	T_{s,N_2}	$1.672\,12 \cdot 10^{-6}$ K
	T_{s,H_2O}	$1.672\,12 \cdot 10^{-6}$ K
Liquid		
Density	ρ_l	1000 kg m^{-3}
Heat Capacity	$C_{v,l}$	4186 J kg^{-1} K^{-1}
Heat of Evaporation	Δh_l^{LV}	$2.5 \cdot 10^6$ J kg^{-1}
Diameter	$d_{droplet}$	20 µm
Solids Mass Fraction	$x_{s,0}$	0.3 kg kg^{-1}
Injection Rate	$\dot{N}_{droplet}$	$1 \cdot 10^5$ s^{-1}

Tab. 6.3.: Numerical setup used in the demonstration case (Hoffmann, 2016).

Numerics		
Time Step		
CFD	Δt_{CFD}	$4 \cdot 10^{-4}$ s
DEM	Δt_{DEM}	$5 \cdot 10^{-6}$ s
Scaling Factor (Coarse Graining)	δ_{CG}	4

6.4.2. Drying Behavior and Tracked Quantities

The tracked quantities and experimentally determined porosities and results are shown in Table 6.4.

Tab. 6.4.: Tracked quantities, namely droplet impact solids fractions and evaporation times, and experimentally determined shell porosities for the demonstration case (Hoffmann, 2016).

	Evaporation Time t_{evap}			Droplet Concentration $x_{\mathrm{s,imp}}$			Experiment
Name	Mean	STD	Skew	Mean	STD	Skew	Porosity
	$\mu_0(t_{\mathrm{evap}})$	$\mu_1(t_{\mathrm{evap}})$	$\mu_2(t_{\mathrm{evap}})$	$\mu_0(x_{\mathrm{s,imp}})$	$\mu_1(x_{\mathrm{s,imp}})$	$\mu_2(x_{\mathrm{s,imp}})$	ε
	[s]	[s]	[s]	[kg kg^{-1}]	[kg kg^{-1}]	[kg kg^{-1}]	[–]
E10	26.01	13.33	1.27	0.40	0.11	1.89	0.54
E11	33.45	16.53	1.13	0.38	0.09	2.53	0.64
E12	12.12	10.59	1.90	0.47	0.15	0.99	0.46
E13	11.62	9.44	1.98	0.43	0.13	1.39	0.50

The tracked quantities were recorded from $t = 75$ s on, as this is the time it took for the mean particle temperature and the mean particle water loading (Fig. 6.4) to reach the steady-state. While the mean particle temperature took about 75 s to rise to 95% of the terminal temperature, the mean per-parcel liquid mass reached its terminal value after 30 s. The water mass rose fast in the first 5 s of the simulations. This can be attributed to the low initial temperature of the particles that cooled the inflowing air. In turn, this reduced the saturation vapor concentration of the air and reduced evaporation rates to a degree that did not allow for liquid to evaporate to a substantial degree.

The corresponding liquid layer evaporation time distribution is shown in Fig. 6.5. It is observable that an increase in the spray rate reduces the evaporation time by a larger degree in the cases that have lower temperatures (E10-E11) than higher ones (E12-E13). This can be expected given the larger amount of available energy for evaporation as well as the higher saturation concentrations. The solids fractions of droplets impacting the particles (Fig. 6.6) show a similar trend, where higher temperatures and low spray rates favor higher droplet concentrations.

The drying potential (Fig. 6.7b) is a macroscopic quantity that directly depends on the global process conditions. Hoffmann et al. (2015) hypothesized a purely linear relationship between the drying potential and the shell porosity. In comparison, the micro-scale quantities of liquid film evaporation time and the solids concentration of droplets upon impact (Fig. 6.7a) show at least similar qualities of fit, indicating that these may be effectively applied to a larger case. For a closed form expression of a mapping, further data points are advisable to be able to distinguish between the influence of evaporation time and droplet concentration when granulating solutions.

6.5. Summary

In this chapter, the literature of microscopic models relevant for the micro processes in fluidized bed spray granulation is reviewed. Based on these, a simulation approach for the prediction of product properties was presented. A calibration workflow relating the macroscopic outcome of granulation experiments to microscopic tracked quantities in CFD-DEM simulations, instead of resolving the actual micro processes at work, was proposed.

This approach was then applied to an example scenario from literature (Hoffmann et al., 2015), where a salt solution was used. The resulting porosity yielded by the salt solution was well-correlated to the tracked quantities.

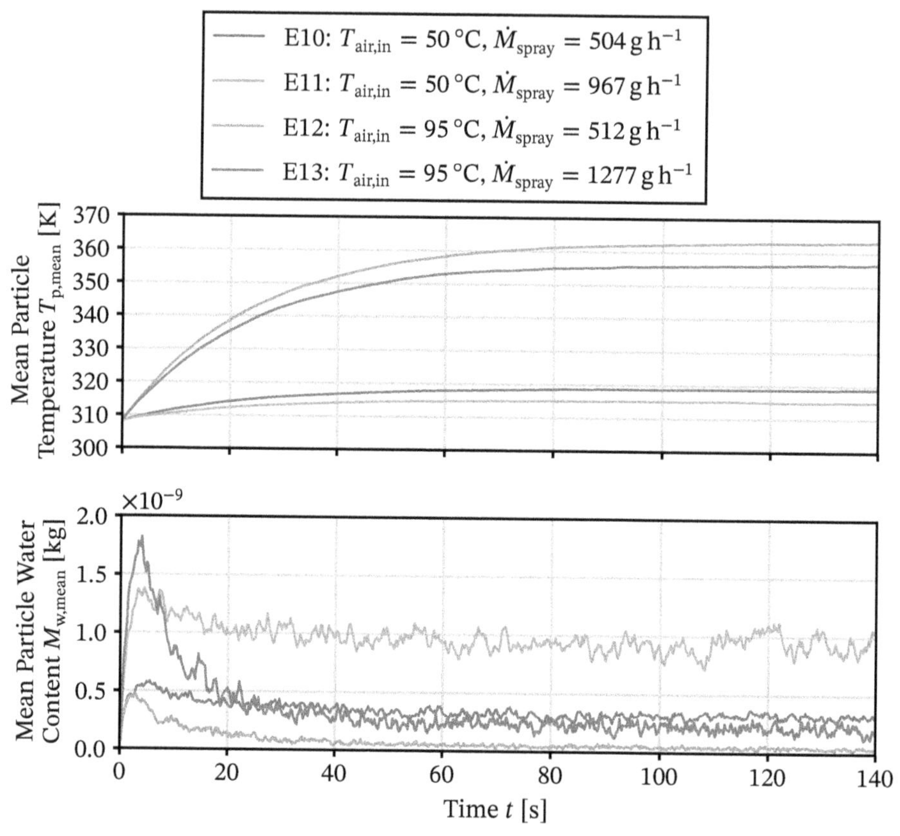

Fig. 6.4.: Mean particle temperature (top) and mean particle water loading (bottom) for the Hoffmann cases.

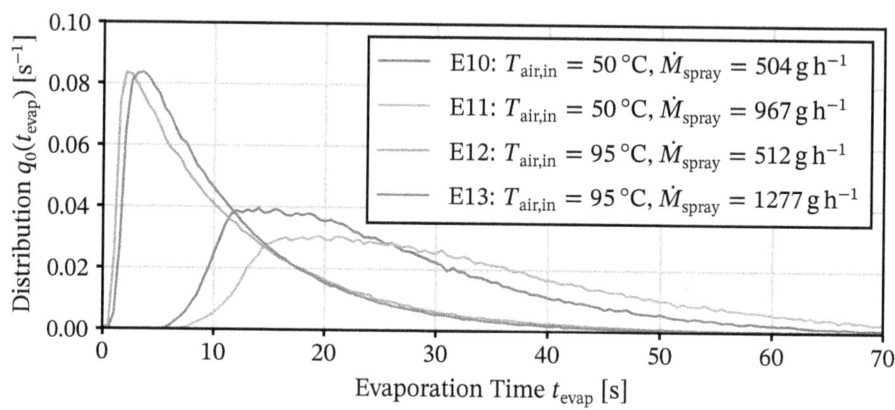

Fig. 6.5.: Particle surface liquid evaporation time distributions for the Hoffmann cases.

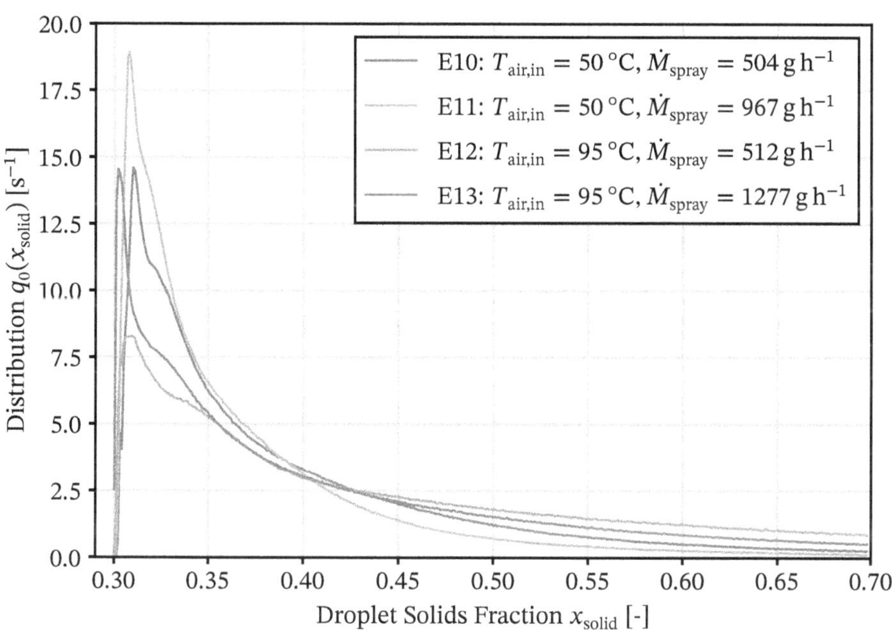

Fig. 6.6.: Solid fraction of droplets impacting the particle surface for the Hoffmann cases.

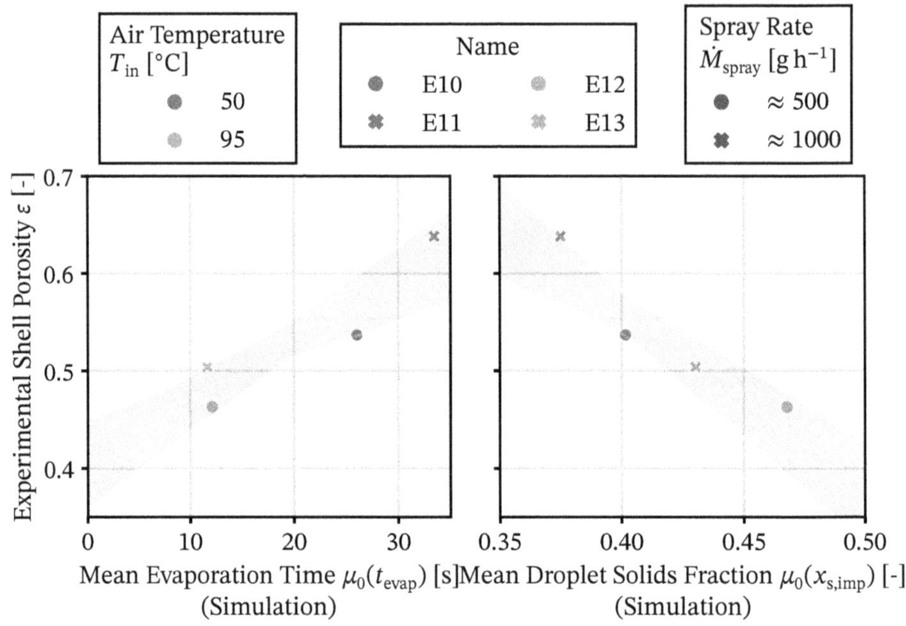

(a) Tracked quantities.

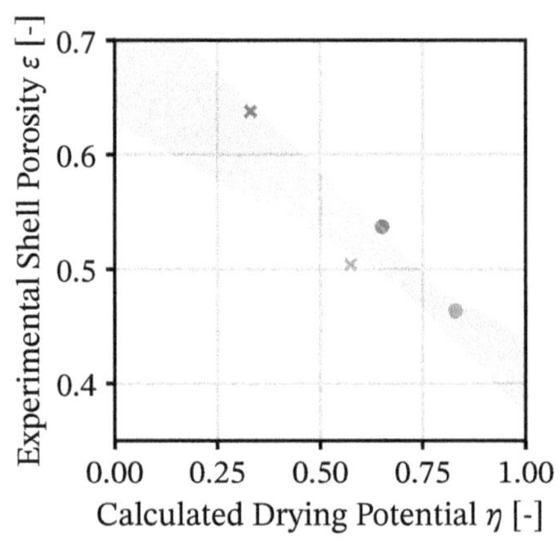

(b) Drying potential.

Fig. 6.7.: Plot of the experimentally measured porosity against two micro-scale, tracked quantities and the macroscopic drying potential in the demonstration case (Hoffmann, 2016). The shaded areas indicate the envelope of all possible linear correlations fitted through the data set.

7 Micro-Scale Product-Property Predictors in Fluidized Bed Spray Granulation of a Suspension

One of the major branches of fluidized bed spray granulation is suspension granulation. Here, a suspension of fine particles in a binder-containing solution is injected into a fluidized bed. The evaporation of the liquid component occurs in flight and after deposition on the surface. Evaporation increases the concentration of the binder and thus the viscosity. As drying progresses, the suspended particles interlock and bound together by the solidifying binder.

The previously described (chapter 6) tracked quantity-product property prediction method was applied to investigate the applicability of the approach to fluidized bed spray granulation that uses a liquid suspension. The cases run in this section correspond to those of Schmidt et al. (2017b). The raw experimental data published there was taken and reevaluated.

7.1. Experimental Procedure

The details of the procedure can be found in the works of Schmidt et al. (2017b). The particles were fluidized in the WSA150 granulator under the same conditions described in chapter 6.4. The varied process conditions are

- the temperature of the inflowing air,
- the spray rate,
- the solids concentration in the spraying suspension and
- the atomization pressure,

equating to changes in droplet size, spray pattern and air flow rate. Schmidt et al. (2017b) calculated the resulting shell porosity exclusively from the shift in the particle size distribution. As particle growth can be highly particle size-dependent, this method is likely less accurate than the one described by Hoffmann et al. (2015), which used X-ray computer tomography. Furthermore, the analysis of dependencies between the shell porosities and process conditions were performed by interpolating over already averaged values rather than using the entirety of populations, which demanded a more rigorous examination of the results.

7.2. Simulation Setup

The geometry of the apparatus was a cylinder with a diameter of 150 mm, as illustrated in Fig. 7.1. The nozzle is setup up in top-spray configuration, 200 mm above the distributor. Because the nozzle used, a *Mod. 970 S4* (Schlick, Germany) was the same as that used by Pietsch et al. (2018a), their approach of adapting the mesh at the nozzle geometry was used. A mesh was generated using the mesher *snappyHexMesh*, shown in the same figure. A few critical regions such as the region with particles, the space close to the nozzles and along the nozzle shaft were refined to generate high-quality cells.

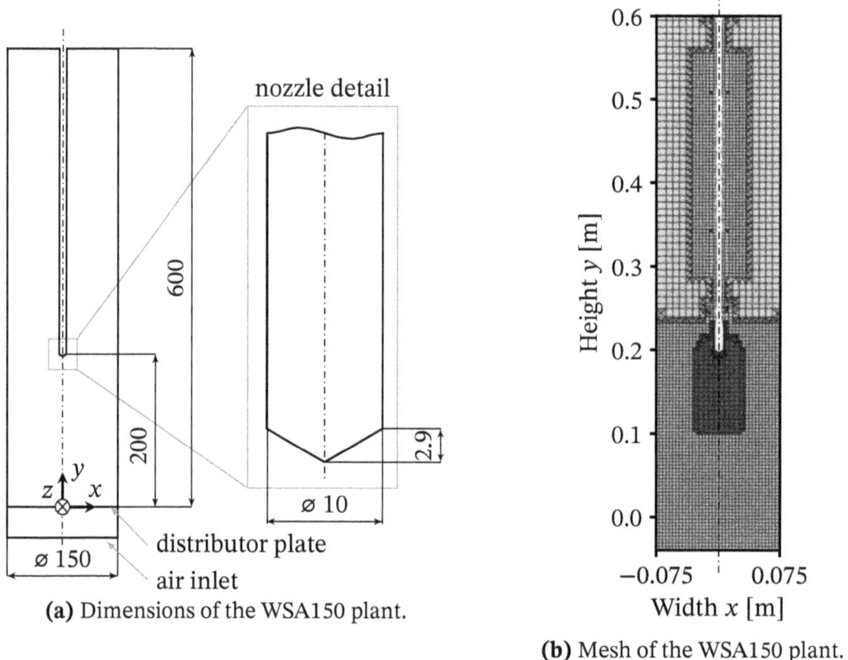

(a) Dimensions of the WSA150 plant.

(b) Mesh of the WSA150 plant.

Fig. 7.1.: Geometry and mesh of the WSA150 granulator.

The particle, spray liquid and gas phase properties as well as the numerical setup are shown in Table 7.1. The spray droplet sizes were estimated to vary with the atomization pressure, shown in Table 7.2. The gas phase was assumed to be an ideal gas that consists of nitrogen gas and water vapor (appendix C.1), with the heat capacity calculated using a mixed approach from the NASA fourth-order temperature-dependent polynomials (appendix C.2) and the viscosity calculated using the Sutherland (1893) equation (appendix C.3). Fluidization air is supplied at 75 kg h^{-1} with an initial bed inventory of 0.5 kg and experiments were performed until 0.23 kg of solids were injected. Simulations were run for a simulation time of 150 s.

Tab. 7.1.: Material properties, contact model parameters and numerical setup of the Schmidt et al. (2017b) case.

Numerics			
Time Step			
CFD	Δt_{CFD}	s	$5 \cdot 10^{-4}$
DEM	Δt_{DEM}	s	$1 \cdot 10^{-4}$
Coupling Interval	$\Delta t_{\mathrm{couple}}$	s	$5 \cdot 10^{-4}$
Scaling Factor (Coarse Graining)	δ_{CG}	-	4
Particle			
Diameter	d_{p}	m	$646 \cdot 10^{-6}$
Density	ρ_{p}	kg m^{-3}	1280
Heat Capacity	C_{p}	J kg^{-1}K^{-1}	840
Young's Modulus	Y_{p}	Pa	$1 \cdot 10^{6}$
	Y_{p}	Pa	$5 \cdot 10^{8}$
Poisson Ratio	η	-	0.22
Restitution Coefficient	e_{pp}	-	0.75
	e_{pw}	-	0.62
Friction Coefficient	$k_{\mathrm{fr,pp}}$	-	0.3
	$k_{\mathrm{fr,pw}}$	-	0.3
Rolling Friction Coefficient	$k_{\mathrm{rfr,pp}}$	-	0.025
	$k_{\mathrm{rfr,pw}}$	-	0.025
Liquid			
Density	ρ_{l}	kg m^{-3}	1000
Heat Capacity	$C_{\mathrm{v,l}}$	J kg^{-1} K^{-1}	4186
Heat of Evaporation	$\Delta h_{\mathrm{l}}^{\mathrm{LV}}$	J kg^{-1}	$2.5 \cdot 10^{6}$
Injection Rate	$\dot{N}_{\mathrm{droplet}}$	s^{-1}	$1 \cdot 10^{5}$

Tab. 7.2.: Assumed nozzle air flow and droplet size resulting from varying the atomization pressure.

Atomization Pressure p_{spray} [bar]	Atomization Air Flow Rate $\dot{M}_{\mathrm{air,noz}}$ [kg h^{-1}]	Droplet Size d_{drop} [µm]
1.8	1.9	42
2.2	2.6	32
3.5	4.2	22

7.3. Relationship between Shell Porosity and Process Conditions

The resulting process conditions are given in Tab. 7.3. No clear relationship between macroscopic drying potential and the shell porosity could be identified (Fig. 7.2a), as described by Schmidt et al. (2017b). Even when considering only varying spray rate and temperature (Fig. 7.2b), no clear relationship can be observed.

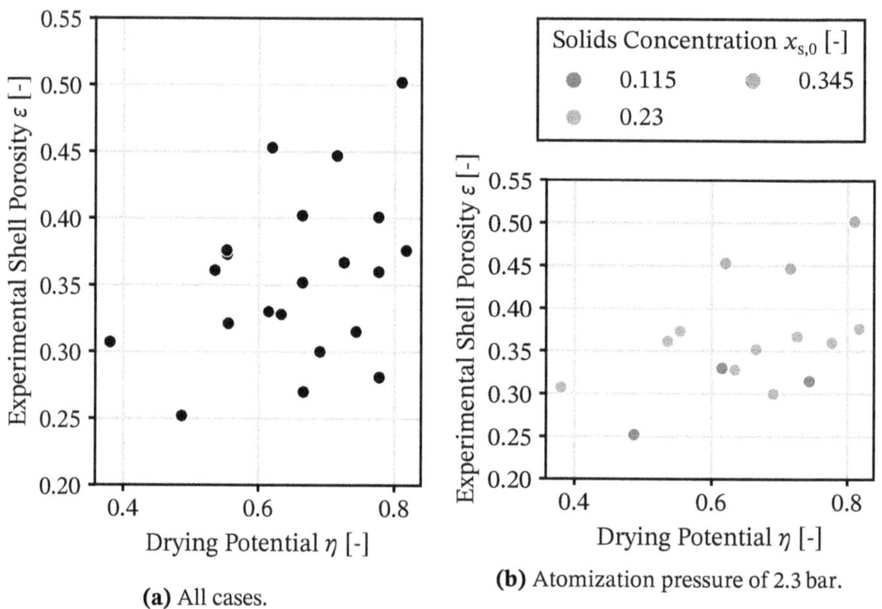

(a) All cases.

(b) Atomization pressure of 2.3 bar.

Fig. 7.2.: Plot of drying potential and experimentally measured porosity. Note that in 7.2b, temperature and concentration are not fixed.

A trend between porosities and the concentrations (Fig. 7.3a) only becomes apparent when considering only a subset at a constant atomization pressure. Here, a higher solids concentration leads to higher shell porosities. Even then, no clear dependence can be identified with the spray rate. This indicates that the atomization pressure, and thus droplet size, plays a central role.

When plotting the dependency of the porosity on the atomization pressure (Fig. 7.4b), no clear relation can be obtained except for the trend that porosity decreases with increasing atomization pressure, possibly owing to smaller droplets being emitted from the nozzle. Controlling for variation in spray liquid solids concentration shows a much clearer dependency (Fig. 7.4b). Here, the porosity decreases with higher atomization pressures. Once again, the spray rate does not show a systematic influence on shell porosity.

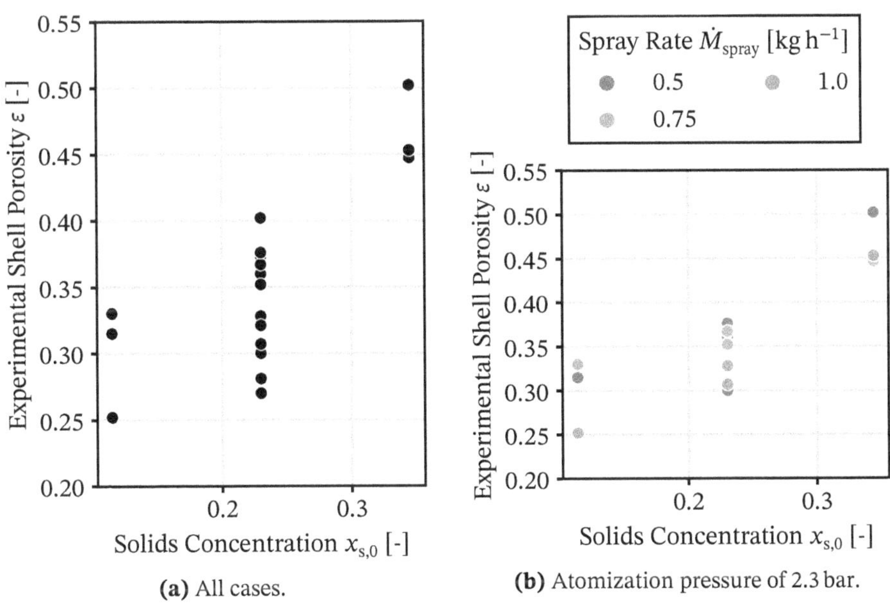

(a) All cases.

(b) Atomization pressure of 2.3 bar.

Fig. 7.3.: Plot of the concentration of solids in the spray liquid and experimentally measured porosity. Note that in 7.3b, the temperature is not fixed

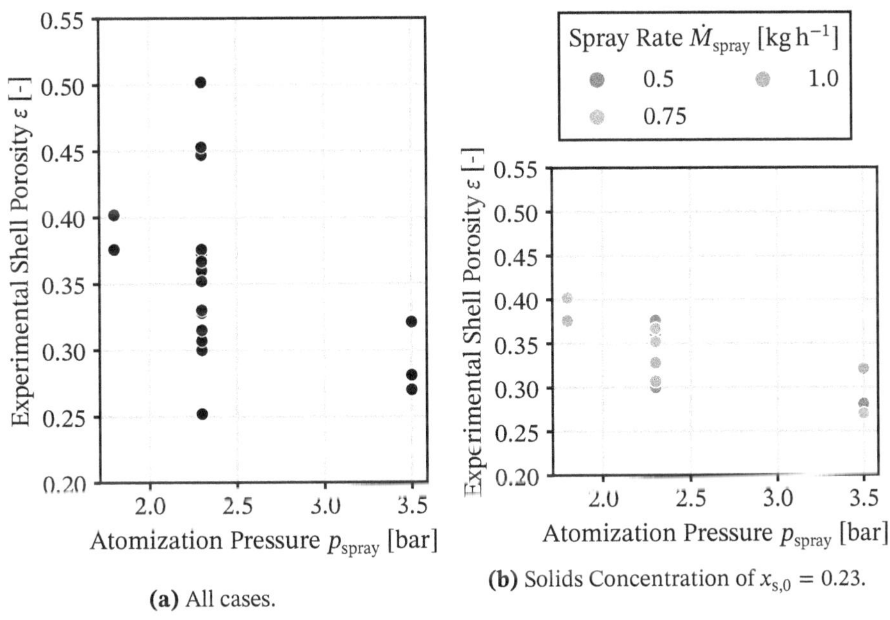

(a) All cases.

(b) Solids Concentration of $x_{s,0} = 0.23$.

Fig. 7.4.: Plot of the spray atomization pressure and experimentally measured porosity. Note that in 7.4b, the temperature is not fixed.

Tab. 7.3.: Process conditions and resulting droplet impact solids fractions, as well as evaporation times and experimentally determined shell porosities for the Schmidt et al. (2017b) case.

	Process Conditions				Evaporation Time		Droplet Concentration		Experiment
Name	Air Temp. [°C]	Spray Rate [$kg\,h^{-1}$]	Solids Fract. [$kg\,kg^{-1}$]	Atom. Pressure [bar]	Mean [s]	STD [s]	Mean [$kg\,kg^{-1}$]	STD [$kg\,kg^{-1}$]	Porosity [–]
A1	60	0.500	0.23	2.3	27.76	14.62	0.27	0.05	0.30
A2	60	0.750	0.23	2.3	37.58	19.35	0.26	0.03	0.36
A3	60	1.000	0.23	2.3	48.04	22.75	0.25	0.02	0.31
A4	80	0.500	0.23	2.3	12.74	7.66	0.28	0.07	0.36
A5	80	0.750	0.23	2.3	14.87	8.30	0.27	0.05	0.35
A6	80	1.000	0.23	2.3	17.66	9.45	0.26	0.04	0.37
A7	95	0.500	0.23	2.3	9.65	6.97	0.29	0.08	0.38
A8	95	0.750	0.23	2.3	10.51	7.11	0.28	0.07	0.37
A9	95	1.000	0.23	2.3	11.74	7.47	0.27	0.06	0.33
B1	80	0.500	0.23	3.5	12.51	7.09	0.30	0.08	0.28
B2	80	0.750	0.23	3.5	16.39	9.15	0.29	0.07	0.27
B3	80	1.000	0.23	3.5	23.13	13.19	0.27	0.06	0.32
B4	80	0.500	0.23	1.8	11.83	7.32	0.28	0.06	0.40
B5	80	0.750	0.23	1.8	13.07	7.66	0.27	0.04	0.40
B6	80	1.000	0.23	1.8	14.60	8.12	0.26	0.04	0.38
C1	80	0.500	0.12	2.3	13.37	7.84	0.14	0.05	0.32
C2	80	0.750	0.12	2.3	16.06	8.80	0.13	0.03	0.33
C3	80	1.000	0.12	2.3	20.32	10.79	0.13	0.02	0.25
C4	80	0.500	0.35	2.3	12.31	7.62	0.42	0.08	0.50
C5	80	0.750	0.35	2.3	13.77	7.94	0.40	0.07	0.45
C6	80	1.000	0.35	2.3	15.84	8.67	0.39	0.06	0.45

Thus, a non-trivial dependency of shell porosity on spray liquid concentration, droplet size, spray pattern and additional heat/mass transfer effects due to more nozzle air can be established. When considering all of the global process conditions simultaneously, a good fit can be achieved:

$$\varepsilon_{\text{shell}} = 0.269 + \begin{pmatrix} T_{\text{air,in}} \\ \dot{m}_{\text{spray}} \\ x_{\text{s,0}} \\ p_{\text{spray}} \end{pmatrix} \cdot \begin{pmatrix} 0.001\,10\,°\text{C}^{-1} \\ -0.0339\,(\text{kg}\,\text{h}^{-1})^{-1} \\ 0.726 \\ -0.0593\,\text{bar}^{-1} \end{pmatrix}. \tag{7.1}$$

The corresponding parity plot, given in Fig. 7.5a, shows fair agreement using this linear approach. However, given the way the shell porosity was quantified, yielding only a single value per experiment, the deviations have to be viewed critically.

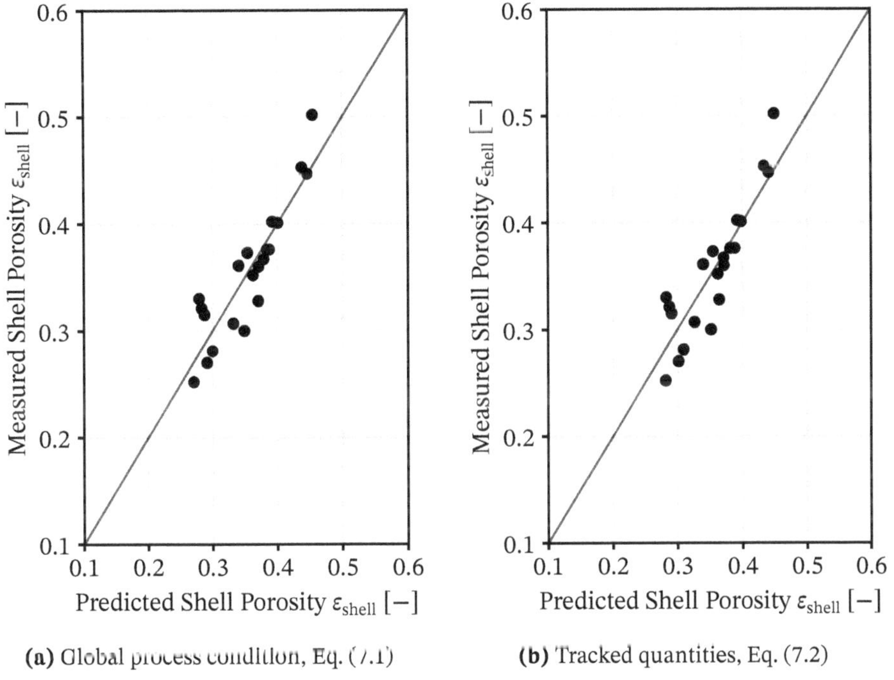

(a) Global process condition, Eq. (7.1)

(b) Tracked quantities, Eq. (7.2)

Fig. 7.5.: Parity plot of the correlations between global process conditions or tracked quantities and the resulting shell porosity.

7.4. Correlating Shell Porosity and Tracked Quantities

The corresponding tracked quantities were, as in the demonstration case (chapter 6.4), the time it took for the liquid film on the particles to evaporate, and the concentration of solids in the droplets at the time of impact.

The particle surface liquid layer evaporation time shows no systematic relationship with the porosity (Fig. 7.6a). The droplet solid concentration at impact (Fig. 7.6b) is only a marginally better predictor for the porosity due to the physics involved - pre-drying of the droplets is not a determinant. Fitting a correlation with feature selection using L_1-regularization yields a three-parameter correlation that only preserves the mean values of the tracked quantities:

$$\varepsilon_{\text{shell}} = 0.301 + \begin{pmatrix} \mu_0(t_{\text{evap}}) \\ \mu_0(x_{\text{s,imp}}) \\ \mu_0(v_{\text{imp}}) \end{pmatrix} \cdot \begin{pmatrix} -0.000\,679 \\ 0.579 \\ -0.0154 \end{pmatrix} \tag{7.2}$$

The corresponding parity plot, shown in Fig. 7.5b, shows very little deviation from the experimental results. Because the three mean tracked quantity values show a strong colinearity with the global process conditions, their parity plots appear very similar (Fig. 7.5a). Yet, due to only having a total of four degrees of freedom compared to five for the process conditions, the tracked quantity correlation in Eq. 7.2 is deemed superior.

Due to the lack of replication experiments, no estimate of the uncertainty involved can be given. Furthermore, lacking knowledge of the droplet size distribution produced by the nozzle at different spray pressures may be at fault for the mediocre collapse of the resulting porosity to the evaporation time and impact concentration - especially since the microscopic process of granulation of suspensions involves the adhesion of the solid-containing droplets due to the viscosity of the binder and its drying.

7.5. Summary

In this chapter, the applicability of the tracked quantity-product property prediction method was applied to the case of fluidized bed spray granulation using a suspension with respect to the porosity formed around the particle cores. Using the drying potential was shown to not be predictive, but performing a linear regression either using global process conditions or tracked quantities performed well in representing the influence of the process conditions. The tracked quantity approach yielded a correlation with fewer parameters, increasing the confidence in the applicability of the method.

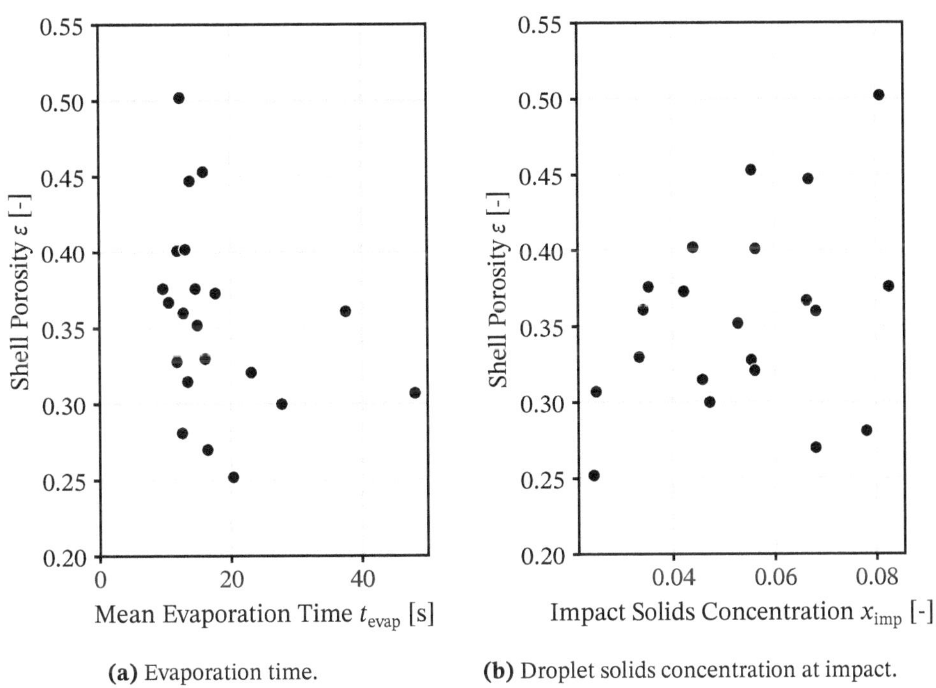

(a) Evaporation time. **(b)** Droplet solids concentration at impact.

Fig. 7.6.: Plot of the particle surface liquid layer evaporation time (Fig. 7.6a) and concentration of the droplets at impact (Fig. 7.6b) against the experimentally measured porosity.

8 Micro-scale Product-Property Predictors in Fluidized Bed Spray Granulation of a Liquid Solution

This chapter is dedicated to the application of the developed method (chapter 6) to a case where a solution of a crystalline solid in a liquid is injected into a fluidized bed. The experimental data from literature that was used in section 6.4 did not cover a wide parameter space that allows for thorough analysis of the applied method. Therefore, similar own experiments under variation of fluidization air temperature, fluidization air flow rate, atomization pressure and atomization air temperature were performed.

To this end, an overview of the granulation experiments is first given, followed by explanations of the roughness quantification method and measures. The simulation setup is explained and the relationship between tracked quantities and the global process conditions is analyzed. This is followed by the derivation of mappings between tracked quantities and product roughness, as well as global process conditions and product roughness. The nature of these mappings is further analyzed in a principal component analysis between product roughness and global process conditions or tracked quantities, respectively.

8.1. Experimental Procedure

Details of the experimental procedure can be found in Orth et al. (2021). Granulation experiments were performed in the Glatt GF3 fluidized bed granulator installed at the Institute of Solids Process Engineering and Particle Technology at the Hamburg University of Technology. It is equipped with a fluidization air heater and a fan that operates in suction mode. A solution of 30 wt% sodium benzoate in distilled water was injected into the process chamber using a peristaltic pump connected to a Schlick 970 S3 two-fluid nozzle (opening diameter 1.2 mm), which is installed in bottom-spray configuration. The atomization air was supplied from the in-house pressure lines and fed through a preheater into the two-fluid nozzle.

2 kg of base particles, Cellets 500, were filled into the process chamber at the start of every experiment and heated up to the process conditions before liquid injection was

started. Liquid injection was terminated after 1 kg of solution, equating to 0.3 kg of sodium benzoate, were fed into the system.

The operating conditions were chosen to vary

- fluidization air flow rate,
- fluidization air temperature,
- liquid spray rate,
- atomization air pressure and
- nozzle air temperature.

The full exploration of this parameter space would result in $3^5 = 243$ combinations of process parameters, equivalent to $243 \cdot 4\,\mathrm{h} = 972\,\mathrm{h}$ of experimental work under the assumption of 4 h of operating time per experiment. As such, a subset of parameters was chosen and three different values were to be evaluated per parameter, as given in Tab. 8.1.

Tab. 8.1.: Different values for parameters in the granulation experiments.

Parameter	Symbol	Unit	Values		
			Low	Mid	High
Fluidization air flowrate	$\dot{V}_g$	$m^3\,h^{-1}$	80	105	130
Fluidization air temperature	T_g	°C	50	85	120
Spray air pressure	p_{noz}	bar	0.5	1.8	3.0
Spray solution flowrate	$\dot{M}_{noz,l}$	$g\,min^{-1}$	10	15	20
Spray air temperature	$T_{noz,g}$	°C	20	70	120

As the focus of this study lies on relating product properties to tracked quantities rather than systematically investigating the physics behind fluidized bed spray granulation, it was chosen to center the experiments around several reference cases, rather than detailing a specific case. The concrete experimental plan, as conducted, is given in Tab. 8.2.

Tab. 8.2.: Experimental plan used for the sodium benzoate on Cellets experiments performed in the GF3 plant at SPE.

	Fluidization air		Spray air		
ID	Flowrate [$Nm^3\,h^{-1}$]	Temperature [°C]	Pressure [bar]	Solution Flowrate [$g\,min^{-1}$]	Temperature [°C]
1	105	85	1.8	10	70
2	130	50	3.0	10	120
3	80	50	0.5	20	20
4	105	85	1.8	15	70
5	105	85	1.8	15	70
6	105	85	1.8	15	70
7	80	50	3.0	10	20
8	105	85	1.8	15	120
9	130	120	3.0	10	20
10	130	85	1.8	15	70
12	80	50	0.5	10	120
13	130	120	0.5	10	120
14	130	50	3.0	20	20
15	130	120	0.5	20	20
16	105	85	1.8	15	20
17	130	50	0.5	20	120
18	80	120	3.0	20	20
19	105	85	1.8	20	70
20	80	120	0.5	20	120
21	130	120	3.0	20	120
22	80	120	0.5	10	20
23	105	85	3.0	15	70
24	130	50	0.5	10	20
25	105	120	1.8	15	70
26	80	120	3.0	10	120
28	105	50	1.8	15	70
30	80	85	1.8	15	70
31	80	50	3.0	20	120
32	105	85	0.5	15	70

8.2. Product Roughness Quantification using Laser-Scanning Confocal Microscopy

To quickly quantify product surface roughness, laser-scanning confocal microscopy was chosen. This method for roughness characterization is well-established in particle technology (Diez et al., 2018; Kato et al., 2006; Pygall et al., 2007) and materials science (Merson et al., 2017). Laser-scanning microscopes operate by tracking a laser-beam across the sample surface. It passes from the laser source through the optical system, including a focusing objective, to the sample. From there, the beam travels back to a half-mirror, where the beam gets redirected through another objective lens and a pinhole onto a photomultiplier. By adjusting the pinhole size, the vertical level $h(\mathbf{x}_i)$ in which a coherent image forms can be found and recorded. The sample area $\{\mathbf{x}_i\}$ can be rapidly scanned by tilting the half-mirror.

8.2.1. Post-Processing of Roughness Measurements

The resulting data provided by the sensor is processed by the software communicating with the microscope. From the 3D profile, as shown in Fig. 8.1a, analysis areas or line profiles can be extracted. These areas and profiles are post-processed using filters to differentiate the roughness of interest from

- the primary profile underlying the roughness, such as curvature,
- sensor noise and
- finer surface features that are not of interest.

To this end, the profiles and areas $h(\mathbf{x})$, where $\mathbf{x}$ is a one or two dimensional space coordinate, are subjected to high and low pass operations that pass a specific range of surface features.

Fig. 8.1b shows the consequence of these operations applied to a line profile that was extracted from the particle surface from Fig. 8.1a. The low pass filter, chosen to be at $\Delta x_{\mathrm{min}} = 8\,\mu\mathrm{m}$, removes the finer features in the length direction from the profile but retains the overall shape. To remove the curvature while retaining the finer features, a high-pass with a cutoff at $\Delta x_{\mathrm{max}} = 80\,\mu\mathrm{m}$ removes the overall curvature while preserving finer features (note that in the plot, the height of the high pass-filtered profile has been shifted to enhance visibility). The combination of high and low pass results in a height profile that contains a selected set of features that can be correlated to specific phenomena, such as dried droplets or growing crystals.

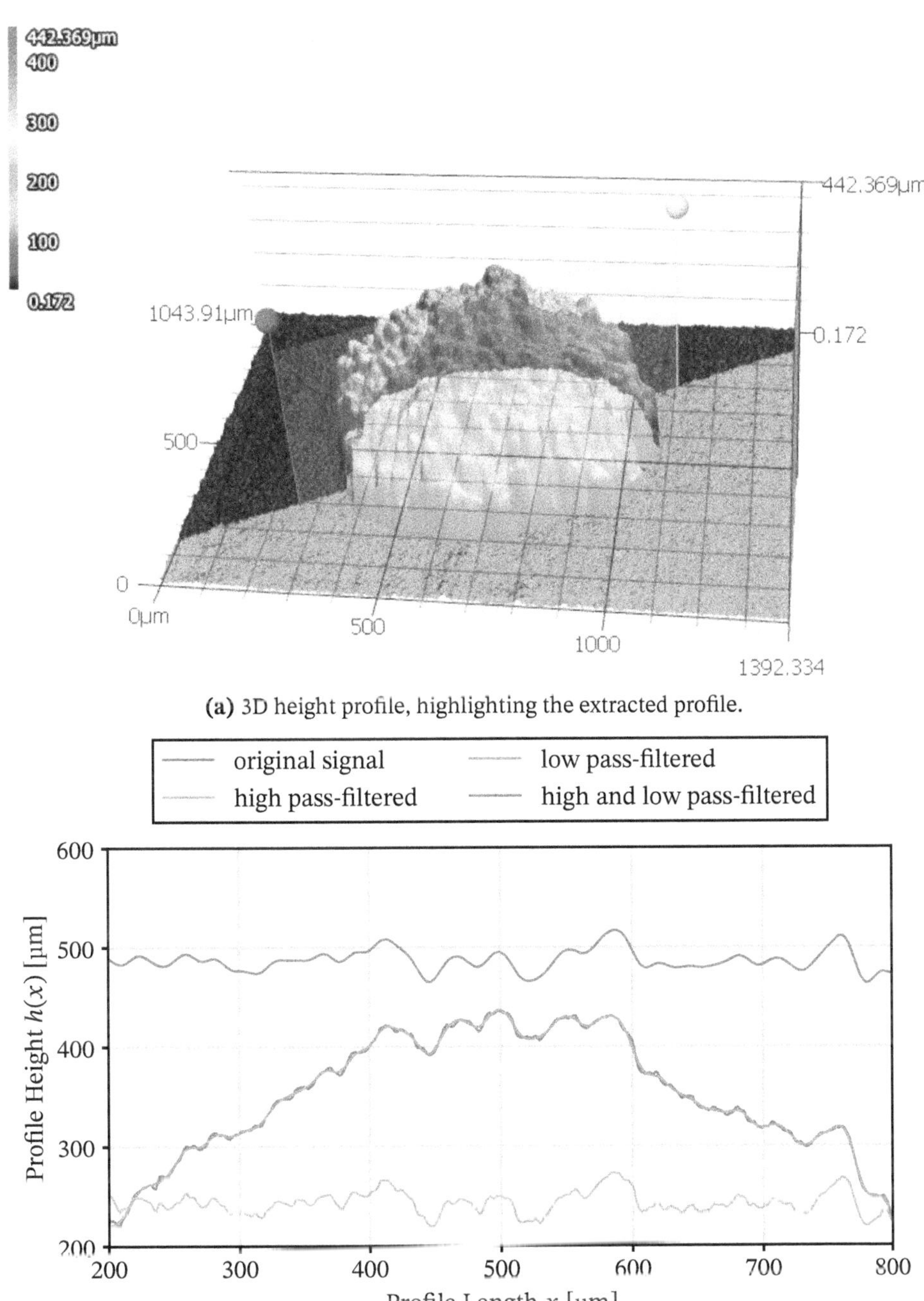

(a) 3D height profile, highlighting the extracted profile.

(b) Original and Processed Height Profiles Profiles.

Fig. 8.1.: Example result of laser-scanning confocal microscopy: 3D profile (Fig. 8.1a) and post processing of extracted line profile for later roughness calculation (Fig. 8.1b).

8.2.2. Roughness Quantification

The agreed standard for surface condition (*DIN EN ISO 4287:1998* 2010) differentiates between roughness, waviness and the primary profile. The roughness R is the part, where only the highly short length scale is considered and the primary profile P the part, where only long length scale features are considered. In the following, the waviness W is technically the target of the analysis as a band pass filter, but the term roughness will be used instead for its colloquial intelligibility.

Roughness can be quantified in a wide variety of ways, depending on the question at hand. The post-processed areas and line profiles $h(\mathbf{x})$ are reduced to scalar quantities. A wide range of quantifiers are possible, two of which are sketched in Fig. 8.2.

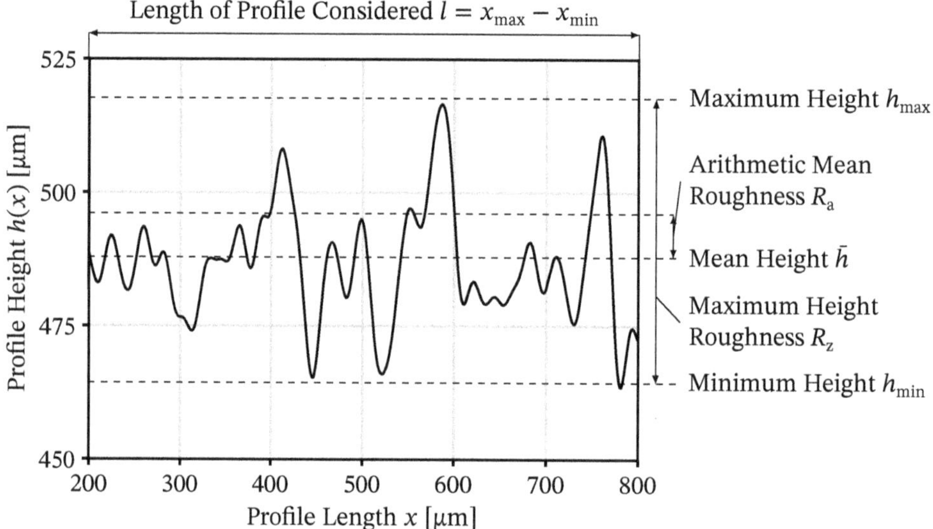

Fig. 8.2.: Sketch of the quantities involved in deriving the arithmetic mean roughness R_a and the maximum profile height R_z in the pre-filtered line profile from Fig. 8.1b.

The most common quantifier of the line roughness is the arithmetical mean height R_a

$$R_a = \frac{1}{l} \int_l \left| h(x) - \bar{h} \right| dx, \tag{8.1}$$

where l is the length of the profile and $\bar{h} = \frac{1}{l} \int_l h(x) dx$ the average height.

The maximum line height is another common parameter, only counting the largest deviation from the mean:

$$R_z = \max_{x \in l} \left| h(x) - \bar{h} \right| \tag{8.2}$$

A more varied quantifier is the root mean square slope that measures the sum over all local gradients:

$$R_{\Delta q} = \frac{1}{l} \int_l \left(\frac{dh(x)}{dx} \right) dx, \tag{8.3}$$

where $\frac{dh(x)}{dx}$ is calculated using a six-point (seven-point) stencil in the case of a measured, discretized profile:

$$\begin{aligned} \frac{dh(x)}{dx} = \frac{1}{60x} &(h(x_{i+3}) - 9h(x_{i+2}) + 45h(x_{i+1}) \\ &-45h(x_{i-1}) + 9h(x_{i-2}) - h(x_{i-3})) \end{aligned} \tag{8.4}$$

By representing the mean slope, this measure gives large values for profiles derived from surfaces with a large specific interface. This, of course, has implications for wetting behavior of the surface in question.

These parameters have area roughness parameter correspondences, such as the arithmetical mean surface height:

$$S_{\mathrm{a}} = \frac{1}{A} \iint_A \left| h(\mathbf{x}) - \bar{h} \right| d\mathbf{x}, \tag{8.5}$$

where $\bar{h} = \frac{1}{A} \iint_A h(\mathbf{x}) d\mathbf{x}$ is the base plane height. The maximum surface height

$$S_{\mathrm{z}} = \max_{x \in A} \left| h(x) - \bar{h} \right| \tag{8.6}$$

and root mean square surface gradient

$$S_{\Delta q} = \sqrt{\frac{1}{A} \iint_A \left(\left(\frac{\partial h(\mathbf{x})}{\partial x_1} \right)^2 + \left(\frac{\partial h(\mathbf{x})}{\partial x_2} \right)^2 \right) d\mathbf{x}} \tag{8.7}$$

are defined analogous to the line profile parameters.

Another kind of roughness quantifier is the value R_{Sm}, the mean line roughness width. This value is defined as the mean width between zero crossings of $h(x) - \bar{h}$ and thus represents features in the horizontal direction. Applied to the question at hand, this could possibly correlate to the degree of droplet spreading that takes place after impact.

The goal of these measurements is the determination of the states of droplets on the surface. For each experiment, three individual particles were imaged from two sides using a confocal laser-scanning microscope, the principle of which is outlined in section 8.2.

The resulting images/height maps were processed in two steps:

1. *extraction of measurement lines/areas*: For the calculation of surface quantities, three circular areas with a diameter of about 150 µm were placed on the measurement plane. For line quantities, two sets of 50 parallel lines were recorded.
2. *high pass & low pass-filtering*: Using a low pass threshold of 80 µm ensured that no finer roughness features than those of droplets spreading on the particle surface are recorded. The corresponding high pass, set at 8 µm, removed the underlying curvature from the scanned surface profile.

The resulting arithmetic mean height roughness values S_a and R_a are given in Fig. 8.3. The surface roughness S_a was higher than the line roughness R_a in most measurements. Experiments 3 and 17, followed by experiment 12, have yielded the roughest product by both measures.

In the measurements performed, roughness parameters correlated with each other. In Fig. 8.4, an overview of the roughness measures' relationship to each other is given. Specifically, the median value for each of the experiment is used to prevent outliers from influencing results. The mean arithmetic heights for line profiles R_a and surfaces S_a (Fig. 8.4a) show an almost linear relationship for values $R_a > 3$ µm, as can be expected given the similar nature of the analysis. The maximum line roughness R_z correlates perfectly linearly with the mean line roughness R_a (Fig. 8.4c). The same applies for the maximum area roughness S_z and the mean surface roughness S_a, although with a steeper slope. When comparing the mean gradient (subscript Δq) to the mean values for the surface (Fig. 8.4f) and line roughness (Fig. 8.4e), linear relationships can be identified as well. All of these behaviors, especially the relationship between R_{Sm} and R_a that resembles a square-root shape, indicate the presence of spherical caps - the dried droplets - on the particle surface. Overall, keeping the kind of sampling consistent - surface area or line-based - is deemed more than the exact choice of roughness measure as the relationship within these sets correlate very well for the given experiments. Going forward, the surface area-based mean roughness S_a is chosen as the roughness quantifier.

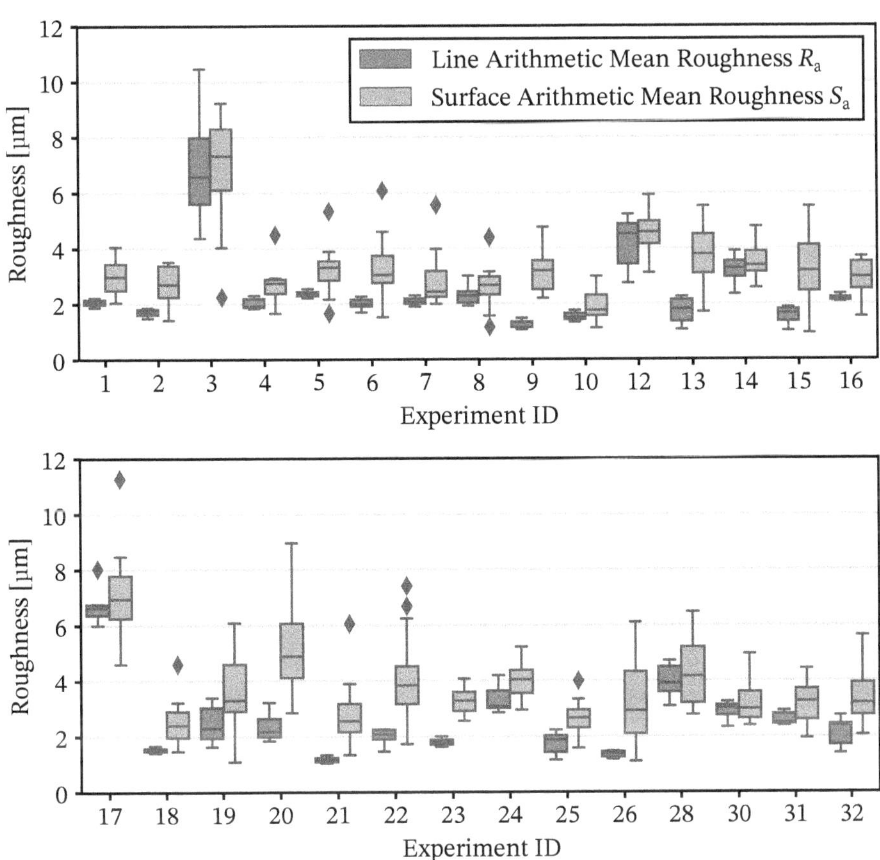

Fig. 8.3.: Overview of the arithmetic mean line roughness R_a and the arithmetic mean surface roughness S_a found in the different experiments. Measurements were performed at 20-fold magnification. Diamond-shaped markers indicate outliers that lie 150% apart from the inter-quartile range. The box indicates the bounds of the quartiles and the whiskers indicate the minimum and maximum value of the outlier-corrected set.

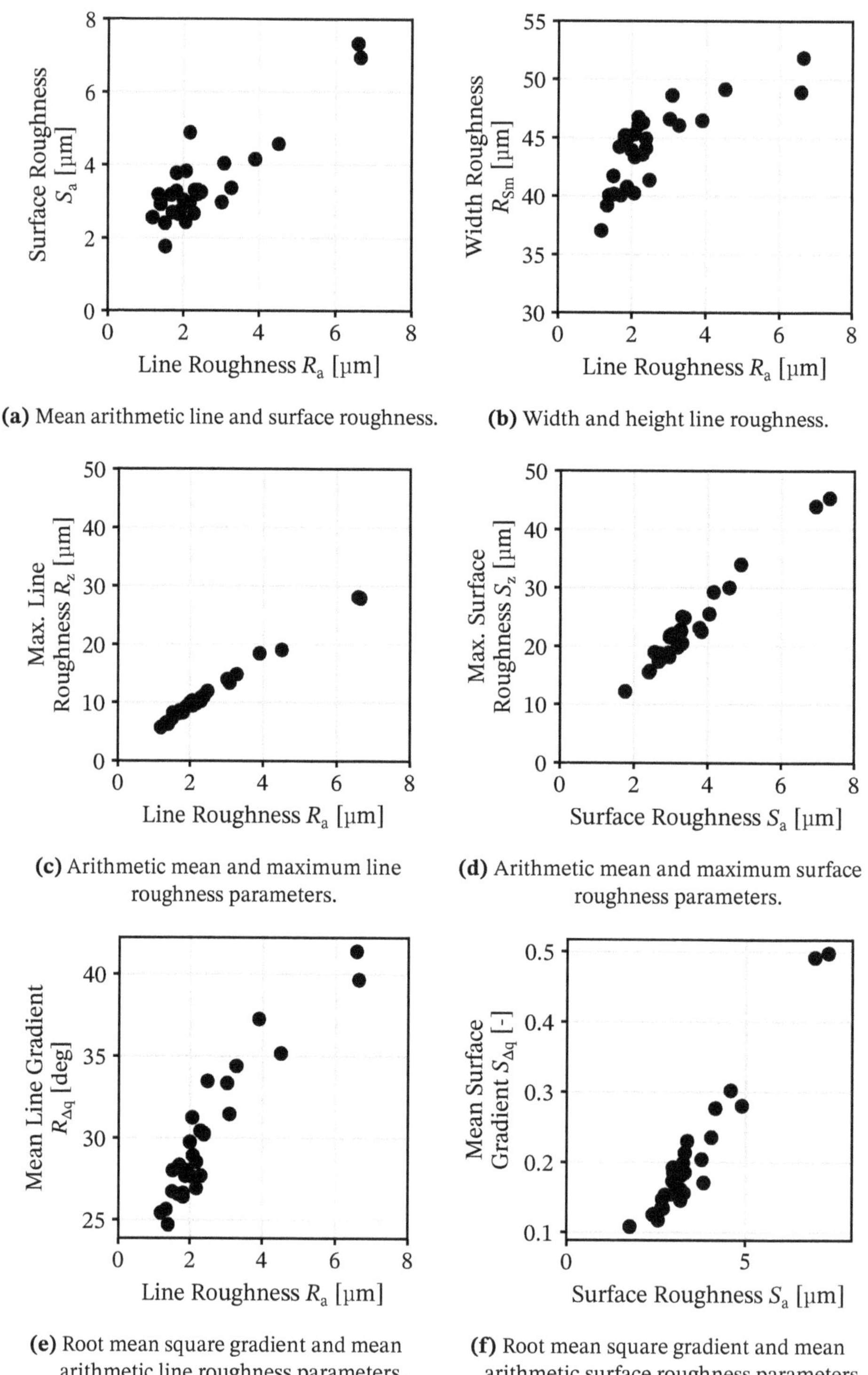

(a) Mean arithmetic line and surface roughness.

(b) Width and height line roughness.

(c) Arithmetic mean and maximum line roughness parameters.

(d) Arithmetic mean and maximum surface roughness parameters.

(e) Root mean square gradient and mean arithmetic line roughness parameters.

(f) Root mean square gradient and mean arithmetic surface roughness parameters.

Fig. 8.4.: Relationship between different kinds of roughness measures.

8.3. Simulation Setup

As in the prior cases, a mesh was created using snappyHexMesh with a grid size of three times the parcel diameter $d_{\mathrm{p,parcel}} = \delta_{\mathrm{CG}} d_{\mathrm{p}} = 2.6\,\mathrm{mm}$ and can be seen in Fig. 8.5a.

At $y = 0\,\mathrm{m}$, a plate was introduced on the DEM side to replicate the effect of a mesh grid that was used in the experiments.

Details of the simulation setup can be found in Table 8.3. Gas phase properties are identical to those given in section 7.2. Droplets are injected above the nozzle patch in order to reproduce the shape of the spray cone. For details on this, please refer to Pietsch et al. (2018b).

Tab. 8.3.: Material properties, contact model parameters and numerical setup of the Orth et al. (2021) case.

Quantity	Symbol	Value
Particle		
Diameter	d_{p}	$650 \cdot 10^{-6}\,\mathrm{m}$
Density	ρ_{p}	$1400\,\mathrm{kg\,m^{-3}}$
Youngs Modulus		
Particle-Particle	Y_{p}	$1 \cdot 10^{6}\,\mathrm{Pa}$
Particle-Wall	Y_{w}	$2 \cdot 10^{6}\,\mathrm{Pa}$
Poisson Ratio	η	0.22
Restitution Coefficient		
Particle-Particle	e_{pp}	0.051
Particle-Wall	e_{pw}	0.051
Friction Coefficient		
Particle-Particle	$k_{\mathrm{fr,pp}}$	0.3
Particle-Wall	$k_{\mathrm{fr,pw}}$	0.3
Rolling Friction Coefficient		
Particle-Particle	$k_{\mathrm{rfr,pp}}$	0.083
Particle-Wall	$k_{\mathrm{rfr,pw}}$	0.028
Scaling Factor	δ_{CG}	4
Time Step	Δt_{DEM}	$1 \cdot 10^{-4}\,\mathrm{s}$
Liquid		
Density	ρ_{l}	$1000\,\mathrm{kg\,m^{-3}}$
Heat Capacity	$C_{\mathrm{v,l}}$	$4186\,\mathrm{J\,kg^{-1}\,K^{-1}}$
Heat of Evaporation	$\Delta h_{\mathrm{l}}^{\mathrm{LV}}$	$2.25 \cdot 10^{6}\,\mathrm{J\,kg^{-1}}$
Droplets per Parcel	N_{droplet}	4

The air flow rates and droplet sizes were determined externally and can be found in Tab. 8.4.

Tab. 8.4.: Measured nozzle air flow and droplet size resulting from varying the atomization pressure.

Atomization Pressure p_{spray} [bar]	Atomization Air Flow Rate $\dot{M}_{air,noz}$ [kg h^{-1}]	Droplet Size d_{drop} [μm]
0.5	2	42
1.8	4	32
3.0	5	22

Spray injection is started at the beginning of the simulation. The particles are suspended in the process chamber to initialize the simulation, with no initial velocity and an initial temperature equivalent to the wet bulb temperature T_{wb} to avoid lengthy equilibration times.

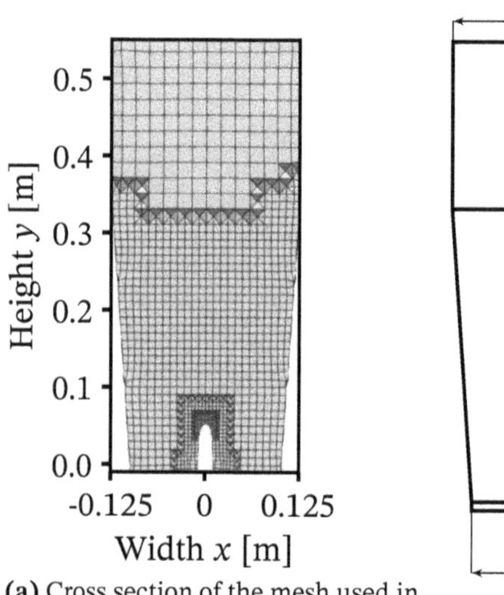

(a) Cross section of the mesh used in the Orth simulation case.

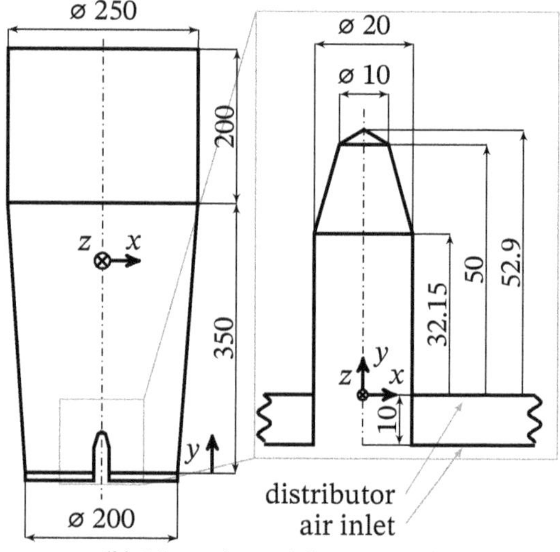

(b) Dimensions of the GF3 model.

For deriving a micro-scale mapping with the surface roughness, the surface liquid evaporation time t_{evap}, concentration in droplets upon impact $x_{s,imp}$ as well as the velocity at which droplets impact the particles' surface v_{imp} are chosen as the tracked quantities. Since recording these quantities creates a large amount of data, only the last 60 s of the simulation are saved for the surface liquid evaporation time and the last 10 s for the other two quantities. The simulation with the highest liquid injection rate and lowest air temperature and flow rate, case 3, was used to estimate the equilibration times required. In this case, liquid accumulation is very slow. This in turn influences the extent to which the mean particle temperature reaches an equilibrium, owing to the spatial unevenness of drying and heating zones and transfer of particles between them by mixing. The mean particle temperature and the mean particle surface liquid over time from case 3 is given in Fig. 8.6. Here, it can be identified that the mean liquid mass $\bar{M}_w$ reaches a steady value after 20 s and the particle temperature after 30 s. For additional safety, the simulation length was set to 60 s overall, allowing for the systems to spend 60 s equilibrating.

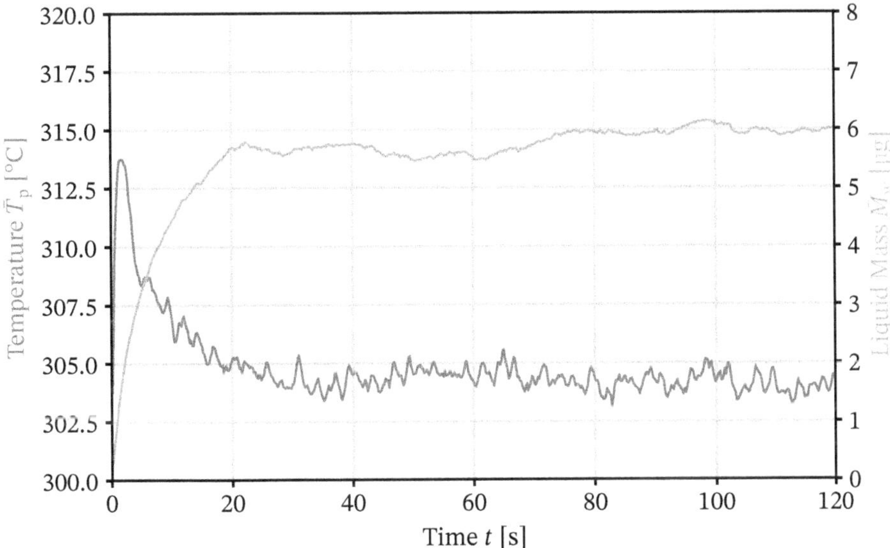

Fig. 8.6.: Mean particle temperature and mean particle surface liquid over time in the case 3 of Orth et al. (2021), with a fluidization gas velocity of $\dot{V}_{air} = 80\,\mathrm{Nm^3\,h^{-1}}$ and temperature of $T_{air,in} = 50\,°\mathrm{C}$, a spray rate of $\dot{M}_{spray} = 20\,\mathrm{g\ min^{-1}}$ an atomization pressure of $p_{spray} = 0.5$ bar and a spray air temperature of 20 °C.

The order of the liquid holdup equilibrating before the mean particle temperature illustrates the aforementioned interplay of wetting, drying and mixing. This is also apparent when observing instantaneous snapshots of the gas temperature, vapor fraction and heat transfer rates in conjunction with particle positions, temperatures and droplet positions in one such case, shown in Fig. 8.7. When focusing in on an area close to the nozzle ($x = 0$ m, $y = 0.05$ m), a region with very few particles due to the high gas velocities can be identified. The gas velocity in the proximity of the particles leads to evaporation of liquid in the droplets, observable by the high vapor mass fraction and low temperatures there. Above this vacuoule, particles directly exposed to the stream of droplets can be observed ($y = 0.15$ m) where gas and particle temperatures are equilibrated and no substantial heat transfer can be seen. This is due to the high water vapor content in this region.

This pattern repeats at the apparatus walls ($x = -0.1\,\mathrm{m}$, $y = 0.05\,\mathrm{m}$), where wetted particles move downwards after leaving the proximity of the nozzle due to the circulation pattern inherent in fluidized beds. The highest heat transfer rates occur close to the nozzle shaft where particles rise up towards the wetting zone. Here, peak local heat transfer rates of up to $400\,\mathrm{MW\,m^{-3}}$ can occur due to unconsumed/uncooled gas preferably rising with bubbles towards the center of the granulator, allowing for large temperature gradients. Due to by-pass in the form of bubbles and equilibrated zones, the net heat transfer rate will be much lower.

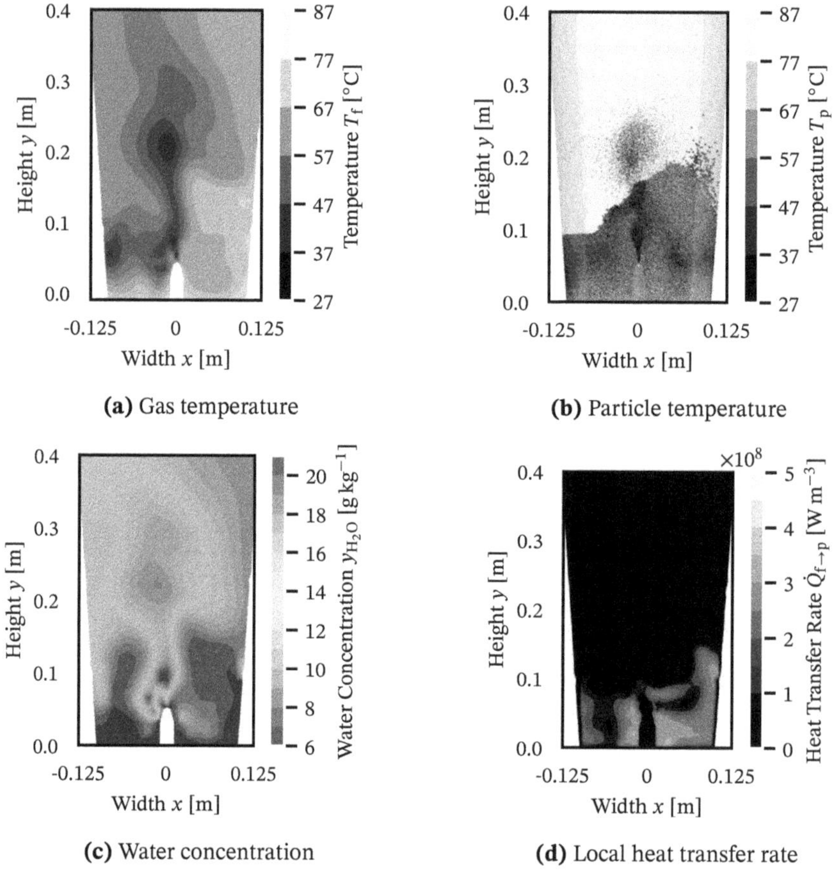

Fig. 8.7.: Instantaneous simulation snapshots showing gas and particle temperature, spray cloud, water concentration in the gas phase and the rate of heat transfer in the apparatus midplane ($z = 0$). The process conditions were set for the reference case at a fluidization gas velocity $\dot{V}_{\mathrm{air}} = 105\,\mathrm{Nm^3\,h^{-1}}$ and temperature $T_{\mathrm{air,in}} = 85\,°\mathrm{C}$, an atomization pressure of $p_{\mathrm{spray}} = 1.8\,\mathrm{bar}$ and a spray air temperature of $70\,°\mathrm{C}$, corresponding to experiments 4-6.

8.4. Relationship between Global Process Conditions and Tracked Quantities

The tracked quantities, namely the surface liquid evaporation time t_{evap}, the solids concentration in droplets upon impact $x_{s,imp}$ and the impact velocity of droplets on the particle surface v_{imp} are the consequence of the global process conditions and the geometry of the apparatus.

All other parameters being equal, variation of the spray rate (Fig. 8.8) does not substantially increase the droplet impact solids concentration distribution. The solids concentration in impacting droplets is determined by droplet drying, which is limited by the saturation of the atomization air. The scenario with relatively low spray rate ($\dot{M}_{spray} = 10\,\text{g min}^{-1}$) thus shows the droplet impact concentration curve shifted to much higher values compared to the other cases ($\dot{M}_{spray} > 10\,\text{g min}^{-1}$). In contrast to this, the effect of increasing the spray rate to a degree in which drying takes place all over the granulator would impact surface liquid evaporation times, while the droplet impact concentration remains the same. This scenario is very much a hypothetical as for this temperature, spray rates higher than $20\,\text{g min}^{-1}$ would suffer from substantial agglomeration.

This exact effect can be seen when considering a variation in fluidization air temperatures (Fig. 8.9). A low air temperature leads to a shift towards longer surface liquid evaporation times and lower solids concentration on particle impact due to lower drying rates and faster saturation of air in the granulator. The droplet impact concentration curves for $T_{air,in} = 50\,°\text{C}$ reveal almost no droplet drying at this temperature, and the two higher temperatures have much wider distributions, albeit with a similar maximum value.

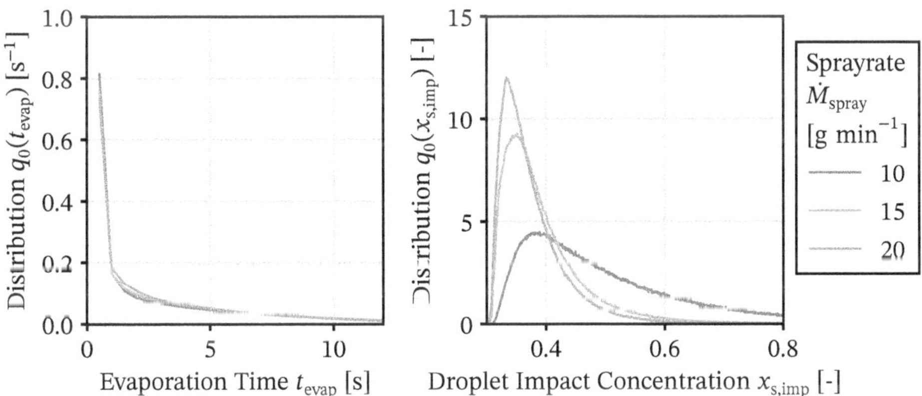

Fig. 8.8.: Influence of spray rate $\dot{M}_{spray}$ on the tracked quantity distributions for the Orth et al. (2021) case at ($\dot{V}_{air} = 105\,\text{m}^3\,\text{h}^{-1}$, $T_{air,in} = 85\,°\text{C}$, $p_{spray} = 1.8\,\text{bar}$, $\dot{M}_{spray} = 15\,\text{g min}^{-1}$, $T_{spray} = 70\,°\text{C}$) unless otherwise indicated.

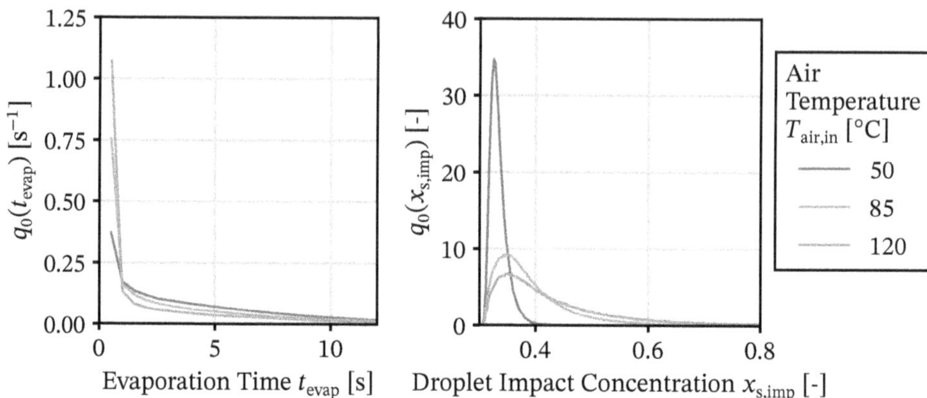

Fig. 8.9.: Influence of air temperature $T_{air,in}$ on the tracked quantity distributions for the Orth et al. (2021) case at ($\dot{V}_{air} = 105\,m^3\,h^{-1}$, $T_{air,in} = 85\,°C$, $p_{spray} = 1.8\,bar$, $\dot{M}_{spray} = 15\,g\,min^{-1}$, $T_{spray} = 70\,°C$) unless otherwise indicated.

8.5. Principal Component Analysis of Influence of Global Process Conditions and Tracked Quantities on Product Properties

Using a linear model has the clear advantage of easy interpretation, but carries with it a penalty in ignoring covariance. To expose colinearities, a principal component decomposition / principal component analysis (PCA) can be used. This method decomposes the dataset of a matrix containing both the product property quantifiers as well as the statistical moments for all process conditions a, referred to as $\mathbf{D}_a = \{\mathbf{M}|y_a\}_a$, into a set of principal components. These components are new vectors that approximate $\mathbf{D}_a$, while remaining both uncorrelated to one another and successively maximizing variance.

Given a matrix $\mathbf{A} = \{\mathbf{a}_i\}_i$, containing column vectors $\mathbf{a}_i$ that represent the values of a single kind of observation - either product property quantifiers or statistical moments - shifted to yield a mean of zero, the corresponding covariance matrix is given by $\mathrm{Cov}(\mathbf{A}) = \mathbf{A}\mathbf{A}^\top$. The covariance matrix is symmetric positive definite and contains the variance of any single kind of observation on the diagonal. The values of off-diagonal entries indicate the degree to which two observations correlate: Small values indicate no correlation and high values a strong correlation.

Performing an eigenvalue decomposition on the covariance matrix gives pairs of eigenvectors $\mathbf{v}^i$ and their corresponding eigenvalues λ_i:

$$\mathrm{Cov}(\mathbf{A})\mathbf{v}^i = \lambda_i \mathbf{v}^i \tag{8.8}$$

Here, a high eigenvalue corresponds to a high variance in the original dataset with respect to the direction of its corresponding eigenvector. The eigenvectors can be used as a basis $\mathbf{V}$ for the dataset $\mathbf{A}$ to yield $\hat{\mathbf{A}} = \mathbf{V}^{\top}\mathbf{A}\mathbf{V}$. When dropping the eigenvectors that correspond to the q lowest eigenvalues, one effectively reduces the dimensionality in a way that minimizes loss of describable variance (Pedregosa et al., 2011).

The result of applying a PCA to global process conditions and the arithmetic surface roughness S_a are given in Table. 8.5.

Tab. 8.5.: Principal components and eigenvalues of the surface roughness and global process conditions.

		Fluidization Air		Spray Air		
Eigenvalue	S_a	Flowrate	Temperature	Pressure	Sln. Flowrate	Temp.
581	0.7	−0.11	−0.35	−0.53	0.31	0.04
402	−0.04	0.61	−0.45	0.41	0.46	0.21
372	−0.02	−0.23	−0.24	0.25	0.25	−0.87
347	0	0.74	0.2	−0.43	−0.2	−0.43
343	−0.01	−0.01	0.69	−0.04	0.72	0.01
163	0.71	0.13	0.31	0.55	−0.26	−0.06

Looking at the size of the eigenvalues and contribution that each successively added principal component gives (Fig. 8.10), all but the smallest eigenvalue contribute to explain the covariance. Going from five to six included principal components increases the median residual by 0.02. This is to be expected considering that all but the last eigenvalues are of comparable size (> 340), while the smallest is less than half of that.

When applying the method to tracked quantities (Fig. 8.11, Table in appendix B.3), it can be seen that three principal components suffice to achieve a median normalized residual of $> 80\%$.

Even using just the principal component with just the highest eigenvalue gives a median normalized residual of 40%, compared with 20% when doing the same with the global process conditions. This asserts that using tracked quantities does not merely perform better by overfitting due to introducing more variables, but is in itself more suitable to represent the physics at hand.

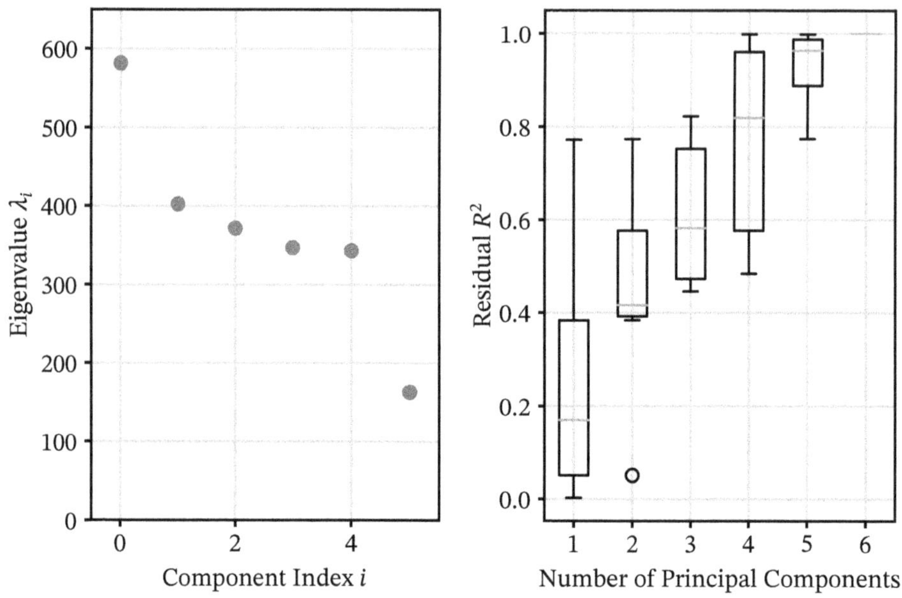

Fig. 8.10.: Size of eigenvalues (left) and distribution of residuals when successively increasing the number of principal components (right) for the PCA of the arithmetic mean surface roughness S_a and the global process conditions.

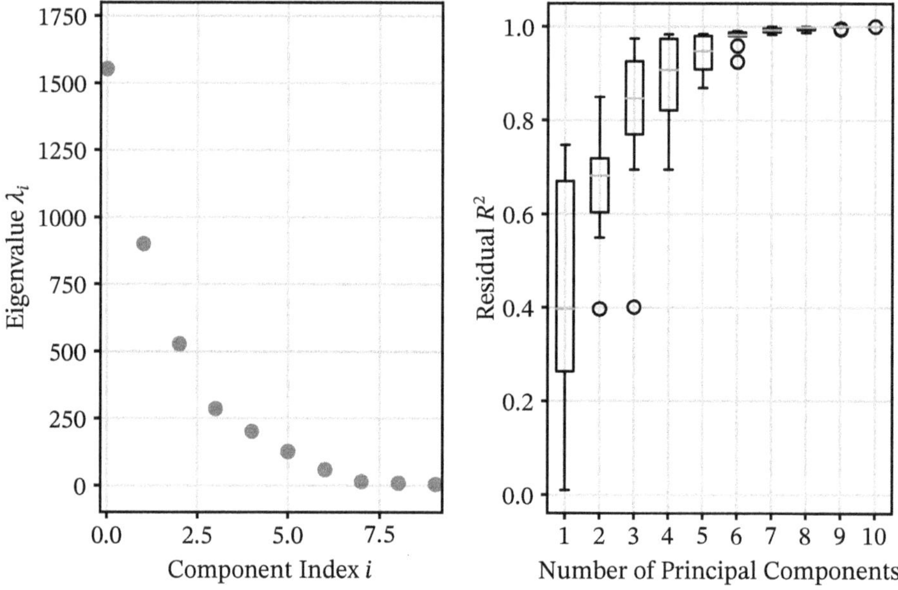

Fig. 8.11.: Size of eigenvalues (left) and distribution of residuals when successively increasing the number of principal components (right) for the PCA of the arithmetic mean surface roughness S_a and the tracked quantities.

8.6. Derivation of a Mapping between Global Process Conditions and Particle Properties

As described in 6.3.1, a mapping between the set of global process conditions varied in the experiments and the median of the mean arithmetic surface roughness S_a was derived by using an ordinary linear regression, yielding:

$$S_a = 5.62\,\mu\text{m} + \begin{pmatrix} T_{\text{air,in}} \\ \dot{m}_{\text{spray}} \\ \dot{V}_{\text{air,in}} \\ T_{\text{atom.,in}} \\ p_{\text{atom.}} \end{pmatrix} \cdot \begin{pmatrix} -0.0154\,\mu\text{m}\,°\text{C}^{-1} \\ 0.0910\,\mu\text{m}\,(\text{g}\,\text{min}^{-1})^{-1} \\ -0.008\,46\,\mu\text{m}\,(\text{m}^3\,\text{h}^{-1})^{-1} \\ 0.001\,83\,\mu\text{m}\,°\text{C}^{-1} \\ -0.740\,\mu\text{m}\,\text{bar}^{-1} \end{pmatrix} \tag{8.9}$$

The precision of this is shown in the parity plot (Fig. 8.12a), where the bulk of measurement is accurately described within 1 µm. Performing a fit using L1 regularization eliminates the contribution of the atomization air temperature $T_{\text{atom.,in}}$ and the fluidization air flow rate $\dot{V}_{\text{air,in}}$:

$$S_a = 4.62\,\mu\text{m} + \begin{pmatrix} T_{\text{air,in}} \\ \dot{m}_{\text{spray}} \\ \dot{V}_{\text{air,in}} \\ T_{\text{atom.,in}} \\ p_{\text{atom.}} \end{pmatrix} \cdot \begin{pmatrix} -0.006\,93\,\mu\text{m}\,°\text{C}^{-1} \\ 0.0315\,\mu\text{m}\,(\text{g}\,\text{min}^{-1})^{-1} \\ 0\,\mu\text{m}\,(\text{m}^3\,\text{h}^{-1})^{-1} \\ 0\,\mu\text{m}\,°\text{C}^{-1} \\ -0.498\,\mu\text{m}\,\text{bar}^{-1} \end{pmatrix} \tag{8.10}$$

The corresponding parity plot (8.12b) reveals little deviation by leaving out these parameters. This implies that the actual contribution of these factors is minimal. An F-test confirms the irrelevance of the spray air temperature $T_{\text{atom.,in}}$ with a p-value of $p > 0.33$. Additionally assuming that the coefficient of the fluidization air flow $\dot{V}_{\text{air,in}}$ rate is zero reduces the p-value to $p > 0.05$. This is commonly agreed to be the threshold for significance and the assumption that neither fluidization air flow rate nor atomization air temperature are significant.

Analogously, conducting a mapping using the line roughness R_a yields similar results:

$$R_a = 4.79\,\mu\text{m} + \begin{pmatrix} T_{\text{air,in}} \\ \dot{m}_{\text{spray}} \\ \dot{V}_{\text{air,in}} \\ T_{\text{atom.,in}} \\ p_{\text{atom.}} \end{pmatrix} \cdot \begin{pmatrix} -0.0265\,\mu\text{m}\,°\text{C}^{-1} \\ 0.0611\,\mu\text{m}\,(\text{g}\,\text{min}^{-1})^{-1} \\ 0\,\mu\text{m}\,(\text{m}^3\,\text{h}^{-1})^{-1} \\ 0\,\mu\text{m}\,°\text{C}^{-1} \\ -0.491\,\mu\text{m}\,\text{bar}^{-1} \end{pmatrix} \tag{8.11}$$

The same parameters, fluidization air flow rate and spray air temperature were eliminated in the fit and both intercept and atomization pressure contribution have similar numerical values. The other two coefficients, the one for the fludization air temperature and the spray rate have both increased in absolute terms, although with opposite signs.

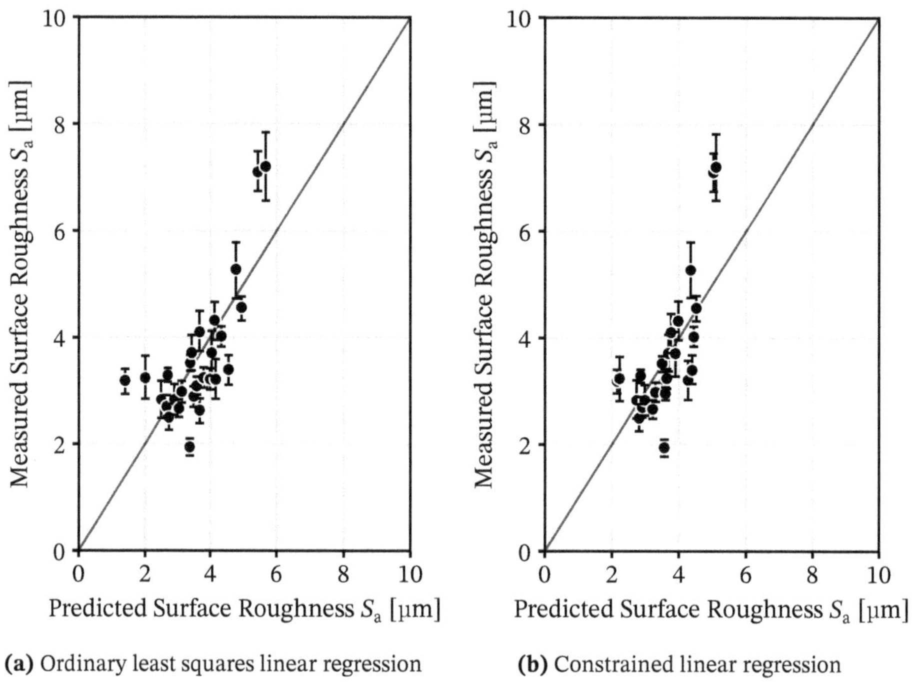

(a) Ordinary least squares linear regression

(b) Constrained linear regression

Fig. 8.12.: Parity plots for the mapping between global process conditions and arithmetic mean surface roughness S_a.

8.7. Derivation of a Mapping between Tracked Quantities and Particle Properties

Finding a mapping between tracked quantities and particle properties, chosen to be the arithmetic mean surface roughness S_a, was performed using the ordinary least squares method as well as L_1-regularized linear regression. As before with the ordinary least squares method, statistically insignificant parameters were eliminated to yield

$$S_a = \begin{pmatrix} \mu_0(t_{\text{evap}}) \\ \mu_0(x_{\text{s,imp}}) \\ \mu_1(x_{\text{s,imp}}) \\ \mu_2(x_{\text{s,imp}}) \\ \mu_2(v_{\text{imp}}) \end{pmatrix} \cdot \begin{pmatrix} -0.796 \\ 15.0 \\ -45.8 \\ -1.31 \\ 3.25, \end{pmatrix} \tag{8.12}$$

where units were omitted for brevity and the parity plot can be seen in Fig. 8.13a. Notably, the constant contribution was eliminated entirely (with $p > 0.15$ for this hypothesis), as well as four other tracked quantities to yield a five-parameter expression. To judge the quality of this fit, using the full residual

$$R^2 = \frac{1}{N} \sum_i \left(y_i^{\text{measured}} - y_i^{\text{predicted}} \right)^2 \tag{8.13}$$

is not sufficient, as for every experiment, multiple roughness measurements are present. Therefore, the median roughness $y_i^{50,\text{measured}}$ was used. Thus, the best possible reproduction in a fit would still produce a residual of

$$R^2_{\text{min}} = \frac{1}{N} \sum_i \left(y_i^{50,\text{measured}} - y_i^{\text{predicted}} \right)^2 = 1.32\,\mu\text{m}^2, \tag{8.14}$$

which has to be taken into account. For the given five-parameter expression, a residual of $1.59\,\mu\text{m}^2$ results. In contrast, using a 7-parameter expression that results from both L_1-regularized regression and assuming the relevance of a constant parameter and eliminating other terms results gives the same expression:

$$S_a = 7.91 + \begin{pmatrix} \mu_0(t_{\text{evap}}) \\ \mu_1(t_{\text{evap}}) \\ \mu_1(x_{\text{s,imp}}) \\ \mu_2(x_{\text{s,imp}}) \\ \mu_1(v_{\text{imp}}) \\ \mu_2(v_{\text{imp}}) \end{pmatrix} \cdot \begin{pmatrix} -0.959 \\ 0.493 \\ -29.3 \\ -1.62 \\ -3.44 \\ 2.83 \end{pmatrix} \tag{8.15}$$

This carries a residual of 1.53 µm² and its parity plot is shown in 8.13b. Given the bias of the parameter selection in L_1-regularized regression to include the intercept in the fit on account of no penalty term for this, the equivalence of these is not surprising. On the basis of very similar residuals, the five-parameter expression Eq. (8.12) is preferable. Comparing this to the best fit obtainable using global process conditions, Eq. (8.10) with a residual of 1.97 µm², it can thus be shown that using the tracked quantities is both feasible and, at least in this case, superior to using global process conditions.

Performing the same fit for the line roughness R_a eliminates the same parameters as for the arithmetic-mean surface roughness:

$$R_a = 5.27 + \begin{pmatrix} \mu_0(t_{evap}) \\ \mu_2(t_{evap}) \\ \mu_1(x_{s,imp}) \\ \mu_2(x_{s,imp}) \\ \mu_1(v_{imp}) \\ \mu_2(v_{imp}) \end{pmatrix} \cdot \begin{pmatrix} -0.134 \\ 0.381 \\ -26.5 \\ -0.668 \\ -2.79 \\ 1.61 \end{pmatrix} \tag{8.16}$$

Here, the coefficients and the intercept differ notably from (8.15) in their absolute terms although the signs are identical and the coefficients of the same order of magnitude.

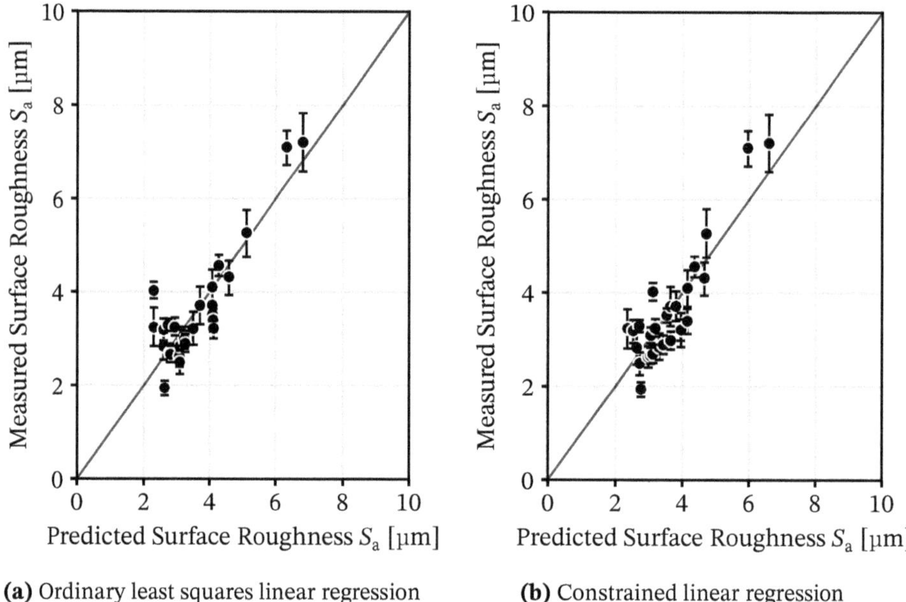

(a) Ordinary least squares linear regression

(b) Constrained linear regression

Fig. 8.13.: Parity plots for the mapping between tracked quantities and arithmetic mean surface roughness S_a.

8.8. Summary

In this section, the applicability of the tracked quantity method to fluidized bed spray granulation using a coating solution was demonstrated. The influence of the process conditions on the product properties as well as on the tracked quantities were studied in detail. Principal component analyses were conducted to eliminate unnecessary complexity in the relationship between global process conditions, tracked quantities and the resulting surface roughness. The method performed remarkably well in fitting the product property data (surface roughness) to tracked quantities.

9 Application of the Product Property Prediction-Approach to Scale-Up

The core issue in scaling fluidized bed granulators is the dissimilarity between mixing patterns that occur on the large scale. These mixing patterns affect the thermal conditions that particles experience, thereby influencing the microprocesses that take place on the small scale and thus determining the properties of the product. Mixing and heat and mass transfer can both be modeled using CFD-DEM, and it was shown (chapter 6) that a mapping approach between the thermodynamic determinants of the microprocesses and product properties has been devised and determined for an example system (chapter 8). The actual application to a scale-up case must now be demonstrated. To this end, an actual scale-up case was considered on the basis of the lab-scale experiment.

9.1. Experimental Design and Simulation Setup

To investigate the ability of the devised method to meaningfully describe the influence of changes in mixing due to scale-up, a scale-up of the base case that was used for the development of the tracked quantity-mapping (chapter 8) was replicated on pilot scale in the *Glatt GF25* plant. The target quantity is the same as used for deriving a mapping, the arithmetic mean surface roughness S_a.

The pilot-scale granulator at hand has four nozzles. By varying the numbers of nozzles in use, the effect of inhomogeneous liquid injection can be emulated. For example, by channeling all of the liquid through a single nozzle, the influence of mixing on the wetting-drying equilibrium can be selectively studied.

The key differences between the pilot and lab scale plants are summarized in Tab. 9.1. The ratio of distributor areas is $0.25\,\mathrm{m}^2/0.0314\,\mathrm{m}^2 \approx 8$. To ensure stability of operation, the fluidization air flow rate $\dot{V}_{\mathrm{air,in}}$ and net spray rates $\dot{M}_{\mathrm{spray}}$ are being scaled linearly with this factor. This way, the drying potential of the air is conserved globally.

The process conditions were chosen to parallel cases 4-6 and 19 of the previous chapter (8). The net spray rate was set at $\dot{M}_{\mathrm{spray}} = 8 \cdot 15\,\mathrm{g\,min^{-1}}$ or $\dot{M}_{\mathrm{spray}} = 8 \cdot 20\,\mathrm{g\,min^{-1}}$

correspondingly. The nozzle configurations that used all four nozzles, the two outermost nozzles and only one nozzle on one end of the fluidized bed were tested in simulations. Because the GF25 plant uses larger nozzles, the cap settings were tweaked until the mean droplet size matched that of the nozzle in the GF3 plant, as measured using laser scattering. Despite this, deviations can be expected due to different spray characteristics.

Tab. 9.1.: Geometric parameters and process conditions used in the scale-up.

	Symbol	Unit	*Glatt GF3*	*Glatt GF25*
Geometry				
Base Dimensions		m	⌀ 0.2	0.25×1
Base Area	A_{in}	m^2	0.0314	0.25
Fluidization Air				
Flow Rate	$\dot{V}_{air,in}$	$Nm^3\,h^{-1}$	105	840
Temperature	$T_{air,in}$	°C	85	85
Spray				
Atomization Pressure	p_{spray}	bar	1.8	1.8
Air Temperature	$T_{atom.,in}$	°C	20	20
Solute Concentration	$x_{s,0}$	$kg\,kg^{-1}$	0.3	0.3
Bed Mass	M_{bed}	kg	2	16

The geometric dimensions of the pilot-scale plant are shown in Fig. 9.1, the adapted nozzle geometry in Fig. 9.2 and the mesh in 9.3. The simulations were conducted with the same settings as described in the previous chapter (8). They were run for 20 s without sampling to establish an equilibrium with respect to both temperatures and moisture content and continued for another 10 s for sampling the tracked quantities.

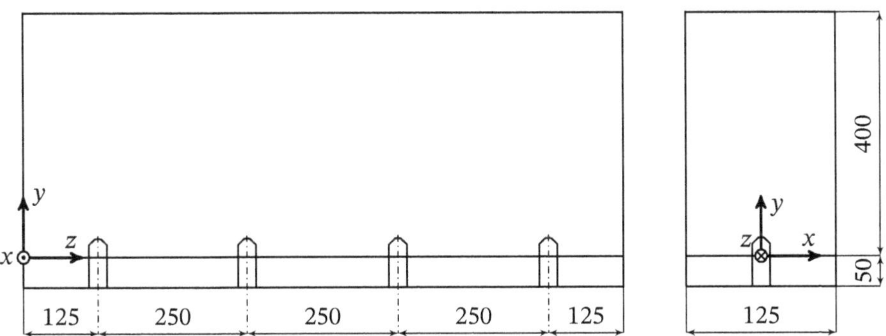

Fig. 9.1.: Dimension of the simulation domain of the GF25 pilot-scale plant.

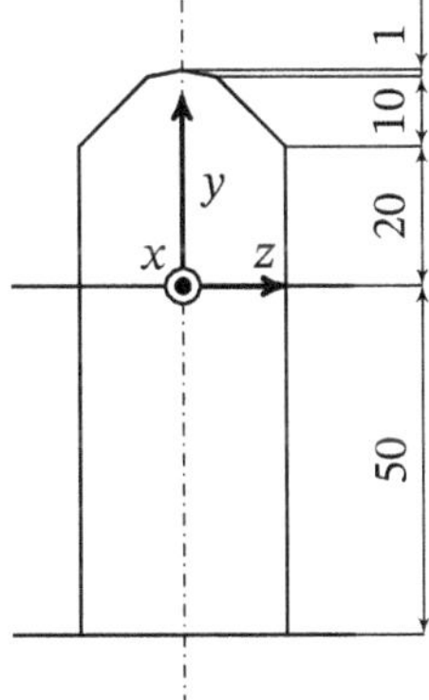

Fig. 9.2.: Dimensions of the nozzle of the GF25 pilot-scale plant.

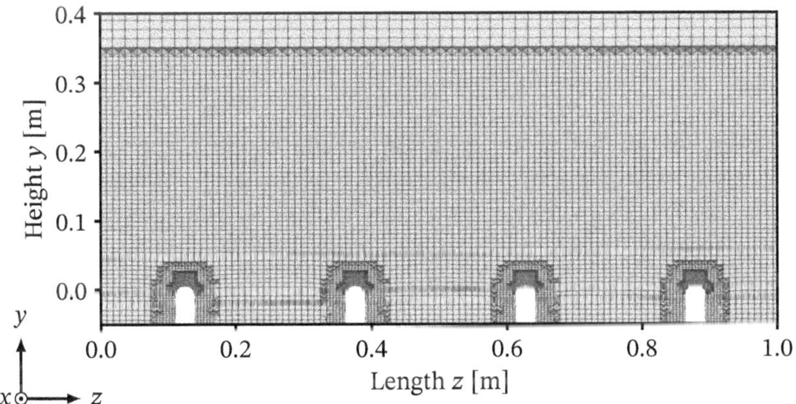

Fig. 9.3.: Cross-section of the mesh used for the simulations of the pilot-scale GF25 plant.

9.2. Analysis of Hydrodynamics, Thermodynamics and Tracked Quantities in Scale-Up

The simulations achieved a stable mean particle temperature after about 8 s for the pilot-scale cases with two and four nozzles, as well as the lab-scale case, while the single nozzle required about 15 s, as shown in Fig. 9.4. All of the simulations approached a long-time mean temperature of 74 °C. A shortcut approximation of the temperature decrease in the system due to evaporation (and neglecting the change in heat capacity of the gaseous phase) gives a temperature change of

$$\Delta T = -\frac{\dot{M}_{\text{spray}}(1 - x_{\text{s},0})\Delta h^{\text{LV}}}{\dot{V}_{\text{air,in}}\rho_{\text{air}}c_{\text{p,air}}} \tag{9.1}$$

$$= -\frac{(120\,\text{g}\,\text{min}^{-1})(1 - 0.3)(2500\,\text{kJ}\,\text{kg}^{-1})}{(840\,\text{m}^3\,\text{h}^{-1})(1.2\,\text{kg}\,\text{m}^{-3})(1.01\,\text{kJ}\,\text{kg}^{-1}\,\text{K}^{-1})} \tag{9.2}$$

$$= -11\,\text{K}. \tag{9.3}$$

Given the inflowing air temperature of 85 °C, this is in line with the resulting particle temperatures, confirming that most of the evaporation takes place on the particles surface rather than by spray in the gas phase. The corresponding per-particle surface liquid content (Fig. 9.4) however, differ drastically between cases. While the case with four nozzles and the lab-scale GF3 case show almost the same temporal behavior, reaching an equilibrium surface liquid content of 1.3 µg within less than 5 s, the equilibrium surface liquid content of the system with two nozzles was at 2.5 µg and that with one nozzle at 6 µg. Given that these figures average over the entire system, only the rate at which they approach equilibrium values show an influence of the position of liquid injection. Furthermore, averaging over the entire apparatus glances over the complexities of mixing within that is taking place.

Looking at the time-averaged spatial distribution of particle surface liquid and temperature along the apparatus length (Fig. 9.5), the reason for the disparity becomes apparent: the particle temperatures are the lowest where liquid is injected and the mean surface liquid is the highest. For the case with four and two active nozzles, an entirely symmetric profile of temperature of particle surface liquid mass can be observed. While for four nozzles, the liquid mass peaks around 2.5 µg at the nozzles and falls down to less than 1 µg between the nozzles, the two nozzle-scenario peaks at above 6 µg and falls down to almost 0 µg for the innermost 0.4 m of the apparatus. With one nozzle in operation, the particle liquid mass reaches 23 µg at the location of injection and falls down to almost zero at a distance of 0.5 m.

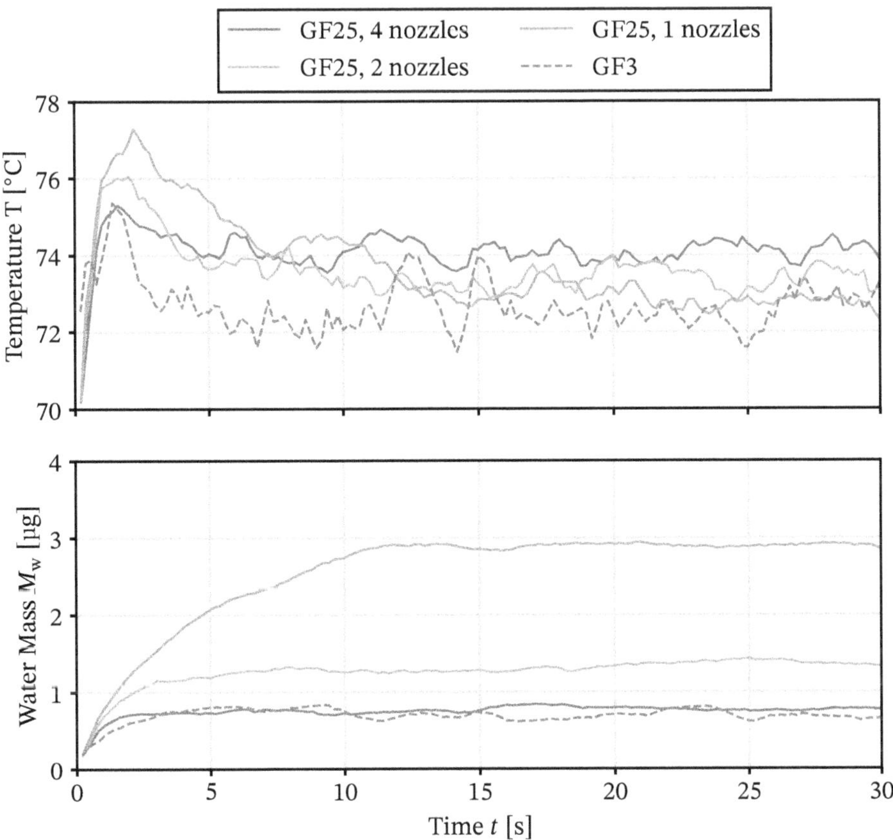

Fig. 9.4.: Mean particle temperature and per-parcel water mass over time for the GF25 pilot-scale plant in three nozzle configurations and the GF3 lab-scale plant operated at a spray rate $\dot{M}_{\text{spray}} = (8\cdot)15\,\text{g min}^{-1}$.

On the thermal side of the matter, the opposite picture emerges. Wherever liquid is injected or transported to, the temperature drops due to evaporation. As the nozzle air is colder than the fluidization air, this introduces dips at nozzle locations that are operated without liquid injection. Gas bypass at the walls, and thus locally increased particle heating, is visible in every case. With four nozzles in operation, the temperature is relatively homogeneous within a 10 °C-corridor around 70 °C, indicating a very consistent distribution of heating of particles that have been cooled due to evaporation. In contrast, with one nozzle, particles that are about 0.4 m away from liquid injection already reach a temperature of 80 °C and the same applies to two nozzles and a distance of 0.25 m.

The corresponding, but instantaneous gas temperature fields are shown in Fig. 9.6. For four nozzles, clear heating and drying zones can be identified. In the first 0.025 m of the apparatus height, the gas temperature reduced from the initial 85 °C by 10 °C or more, indicating that particles are being heated and/or dried. When one or two nozzles are in operation, regions of the apparatus can be identified, where gas just passes the particle bed with no significant heat exchange.

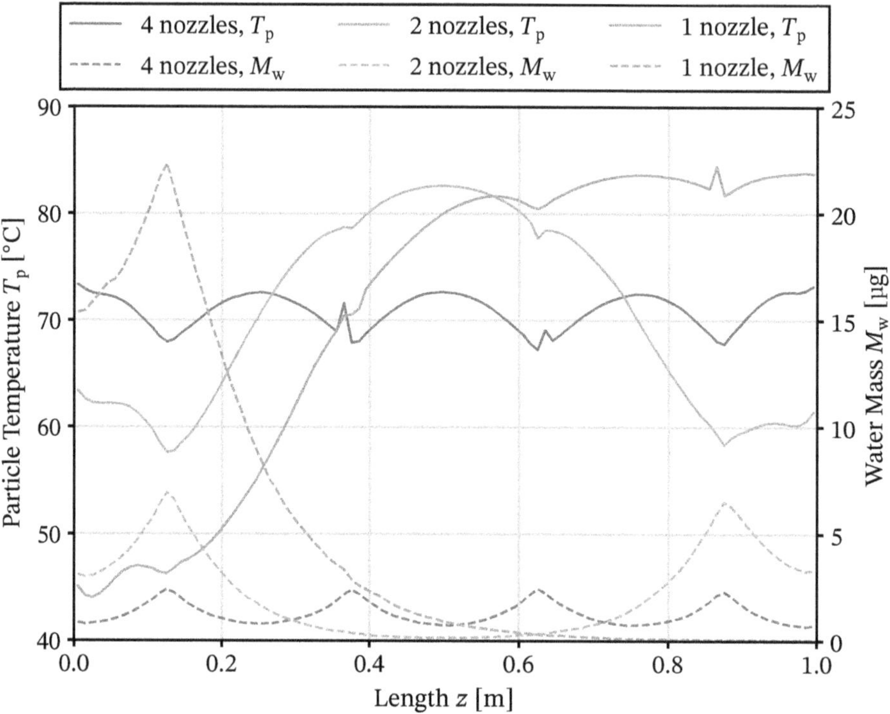

Fig. 9.5.: Distribution of particle surface liquid/water mass and particle temperature across the apparatus length for the same net amount of liquid ($\dot{M}_{spray} = 8 \cdot 15\,\mathrm{g\,min^{-1}}$) injected through four, two and one nozzle.

Due to the data being a cut through the mid-plane where the nozzles operate, very localized spots of cold air are visible that reduce the temperature locally by about 10 °C. All of this shows that for excessive nozzle distances, most of the apparatus operates without contributing to the process at all.

Looking at tracked quantities that result from the process (Fig. 9.7), consequences for the microprocesses on the particle surface can be drawn. The particle liquid layer evaporation time distribution shows that for all cases present, the majority of particles dry in less than 1 s. This can be attributed to the majority of particles that are wetted only by few droplets - the chance of particles to receive a few droplets further away from the spray zone is much higher than to be in the spray zone and receive a lot of liquids.

As the amount of liquid received correlates with the time it takes to evaporate under equivalent conditions, the evaporation time increases correspondingly.

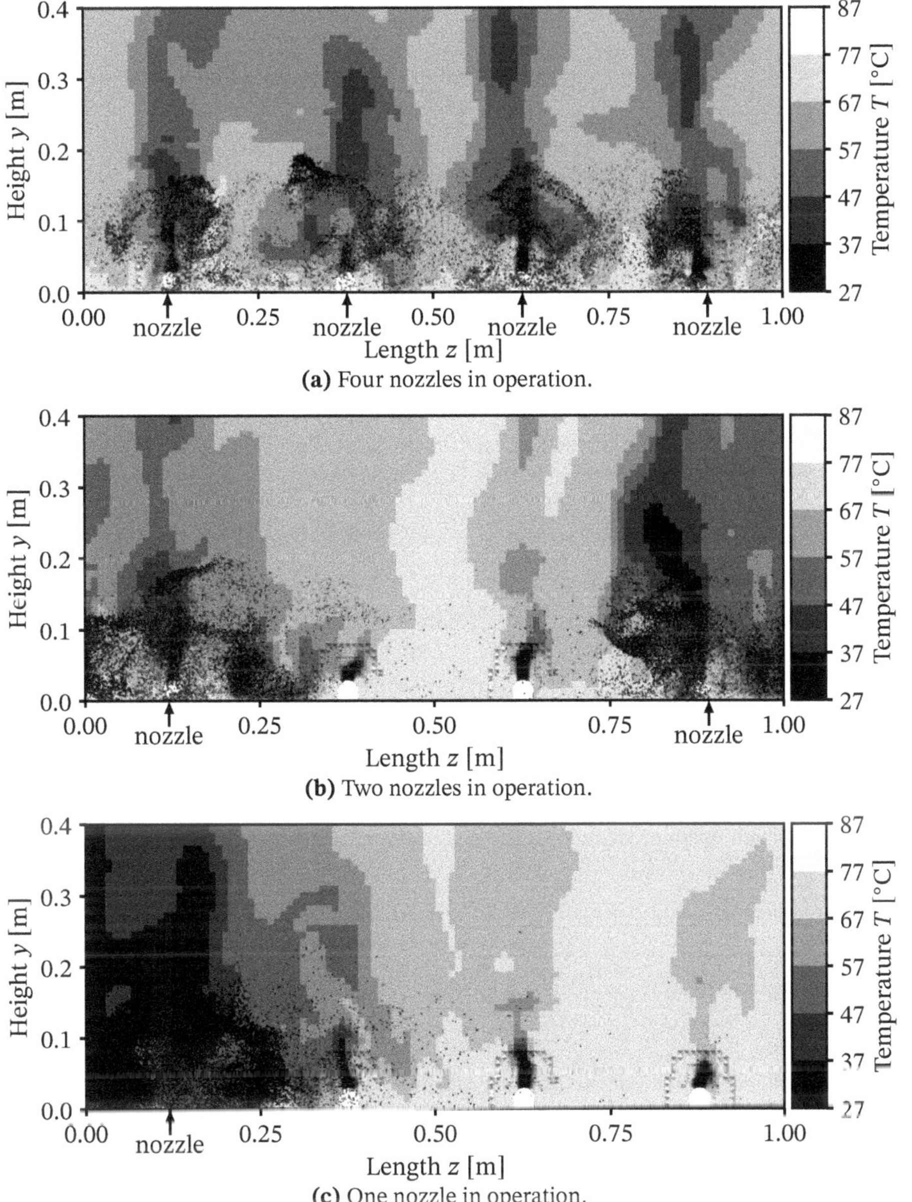

(a) Four nozzles in operation.

(b) Two nozzles in operation.

(c) One nozzle in operation.

Fig. 9.6.: Particle and gas phase temperatures in the GF25 plant (spray rate $\dot{M}_{\mathrm{spray}} = 8 \cdot 20\,\mathrm{g\,min^{-1}}$) after 60 s process duration when introducing the same amount of liquid through four, two or one nozzle. Only wetted particles are shown and colored according to their temperature.

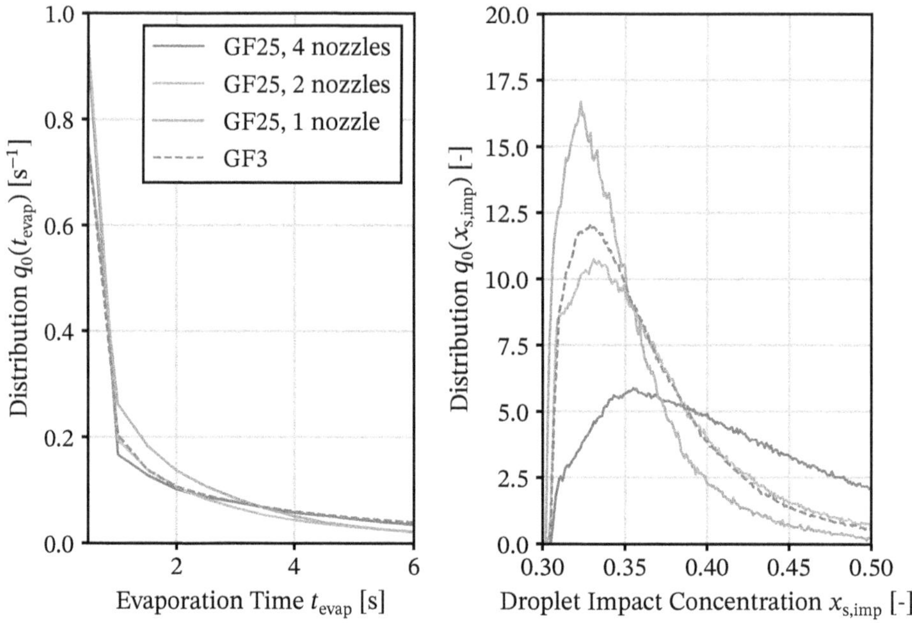

Fig. 9.7.: Tracked quantity distributions at a spray rate of $\dot{M}_{spray} = (8\cdot)15\,g\,min^{-1}$.

The corresponding mean liquid layer evaporation time, shown in Fig. 9.8a as $\mu_0(t_{evap})$, is the highest for one nozzle, followed by two and four nozzles, respectively. Notably, the mean evaporation time on the lab scale (GF3) that corresponds closest to the pilot scale is the one with two nozzles. This is also true for a higher spray rate of $8 \cdot 20\,g\ min^{-1}$, where this trend is much more pronounced: the mean evaporation time with four nozzles lies at 1.8 s whereas the one for two nozzle is at 3.6 s and one single nozzle at 4.9 s. The reason for this behavior lies in the distance that wetted particles have to be transported before drying is completed (as previously mentioned when discussing Fig. 9.5 and Fig. 9.6), as well as the higher amount of liquid per-particle that has to be removed (see Fig. 9.4).

The solids concentration in the droplets upon impact, shown in the same figure, follows a Weibull distribution with a long tail end for all cases. The shape of this distribution can be explained with being a time-average: some droplets are transported further than others and, therefore, have a longer time to dry before encountering a particle.

For the pilot-scale case (GF25) with one nozzle, the peak of the distribution is at a value of $x_{s,imp} = 0.31$, and at $x_{s,imp} = 0.33$ for two nozzles. The four nozzle pilot-scale case has its maximum at $x_{s,imp} = 0.36$. This behavior can be explained with the low rates of drying due to higher humidity of the air.

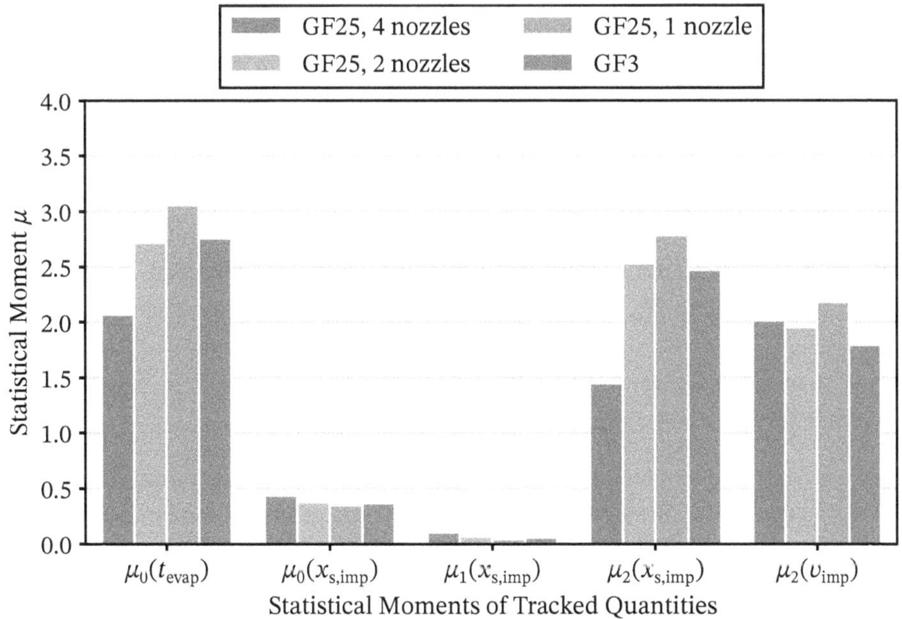

(a) Statistical moments of tracked quantities at a spray rate of $\dot{M}_{spray} = 8 \cdot 15\,\mathrm{g\,min^{-1}}$

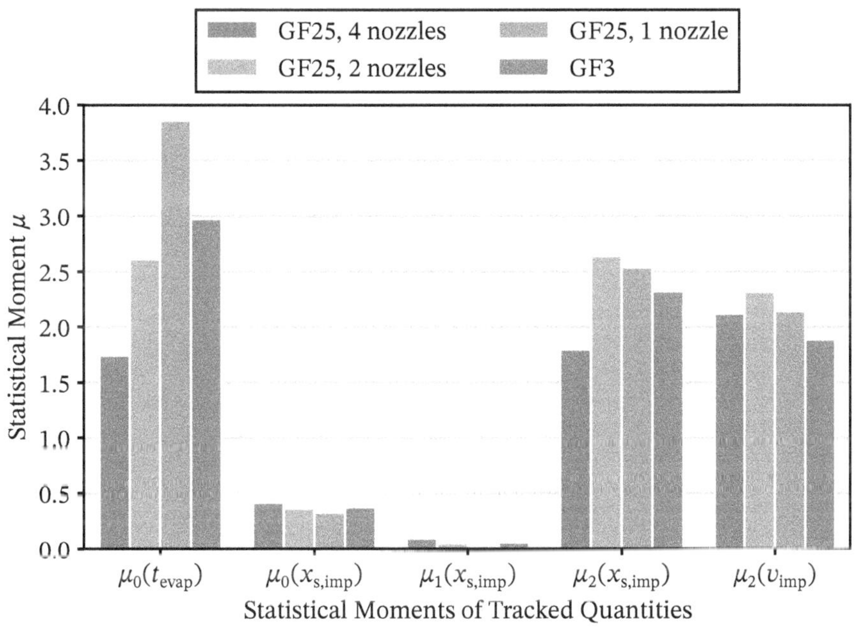

(b) Statistical moments of tracked quantities at a spray rate of $\dot{M}_{spray} = 8 \cdot 20\,\mathrm{g\,min^{-1}}$

Fig. 9.8.: Tracked quantity distributions and their statistical moments for the lab-scale plant (GF3) and pilot-scale plant (GF25).

The degree to which droplets that have a longer lifetime can dry is much more compressed with fewer nozzles, as more liquid is introduced in a smaller area, leading to higher humidity. Interestingly, the solids concentration distribution of the lab-scale apparatus is much more similar to the one obtained at the pilot-scale with one or two nozzle than four nozzles, as is the case with the liquid layer evaporation time distribution.

When considering the statistical moments of the tracked quantities that were identified as the most important in the product property mapping, given in Fig. 9.8a and Fig. 9.8b, the pilot-scale configuration with two nozzles most closely matched that of the lab-scale trial for a spray rate of $\dot{M}_{\text{spray}} = (8\cdot)15\,\text{g}\,\text{min}^{-1}$ with very little deviation. For a higher spray rate of $\dot{M}_{\text{spray}} = (8\cdot)20\,\text{g}\,\text{min}^{-1}$, this is not the case for the second statistical moment of the impact solids concentration $\mu_2(x_{\text{s,imp}})$ and the impact velocity $\mu_2(v_{\text{imp}})$ where the one-nozzle variant is more similar.

When applying the five-parameter mapping between the arithmetic mean surface roughness S_{a} and tracked quantities (Eq. (8.12)), an excellent agreement can be observed in the parity plot (Fig. 9.9).

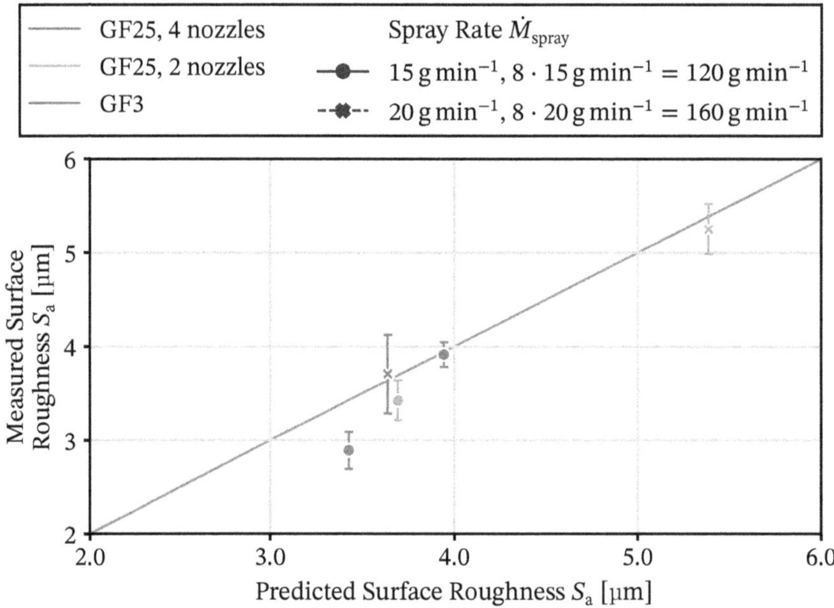

Fig. 9.9.: Parity plot comparing the measured surface roughness data for pilot-scale (GF25) and lab-scale (GF3) granulation experiments and the predicted roughness using the tracked quantity approach.

The case with the lower spray rate, $\dot{M}_{spray} = (8\cdot)15\,\text{g min}^{-1}$, shows a deviation of less than 1 µm with respect to experimentally observed values, with the GF25 with two nozzles producing similar products to the GF3 plant. The GF25 plant with four nozzles on the other hand produces rougher particles. Looking at the predictions, the absolute values are within an accuracy window of smaller than 1 µm. The correct order of roughnesses for the different plant configurations considered can also be described. This degree of accuracy is to be expected given that the shape of distributions of the tracked quantity match.

For a spray rate of $\dot{M}_{spray} = (8\cdot)20\,\text{g min}^{-1}$, the results are spot-on. Impressively, the production of particles with much higher roughness (by 2 µm) for the GF25 plant with two nozzles versus the corresponding GF3 experiment is predicted by the approach. This demonstrates that the approach can accurately depict the deviations in local wetting and drying conditions that arise when scaling to another plant and consider the consequence for these deviations.

9.3. Summary

The developed approach and the mapping derived for the GF3 lab-scale granulator were applied to describe the surface roughness of particles when produced in the GF25 pilot-scale granulator. The nozzle configurations were varied to investigate the ability of the approach to capture the influence of different wetting and drying conditions.

The approach was able to accurately predict the outcome of granulation for all nozzle configurations and spray rates considered, even when the drying conditions differed greatly from the lab-scale trials.

10 Conclusion

In this thesis, two key parts of fluidized bed spray granulation were treated: the phenomenon of liquid bridging that gives rise to agglomeration, and the surface processes that give rise to particle properties. To this end, numerical models were developed that aim to extend the reach of granular simulation science to capture these industrially relevant phenomena.

Depending on the desired product, agglomeration can either be desired or highly problematic in fluidized bed granulators, where liquid on a particles may not instantaneously dry after liquid has been deposited on them. To this end, it is highly advantageous to be able to predict the formation of agglomeration. To describe the dynamics in such weakly wetted granular systems, a novel one-parameter liquid bridge state model for the discrete element method was developed. Its ability to modulate liquid bridge network topology was investigated for a suite of calibration and other liquid bridge contact model parameters. The minimum liquid layer height parameter as well as surface tension were shown to be sufficient to influence the static angle of repose. The dynamic angle of repose additionally depends on the minimum separation distance parameter and liquid viscosity.

The static angle of repose test only showed a systematic response when conducted with a low container filling. In performing the static angle of repose test in simulations, increasing the strength of the liquid bridges by increasing the surface tension was found to be effectively equivalent to reducing the minimum layer height parameter. Low values for the minimum layer height and high values for the surface tension increased the static angle of repose. The experimentally determined static angle of repose for a water-glass system could be replicated in simulations by either artificially inflating the surface tension for low minimum layer heights or using the real value of the surface tension together with high values of the minimum layer height.

The dynamic angle of repose test was found to be more consistent than the static angle of repose test between repetitions of the test. In the numerical model, it yielded the best agreement with the experimental results when using the same parameter set for the surface tension and minimum layer height that performs best in the static angle of repose test. With respect to the parameters relevant for dynamic properties, the liquid viscosity and the minimum separation distance, higher values for the liquid viscosity and lower values for the minimum separation distance increased the dynamic angle of repose. This only applies up to a point, beyond which cohesive forces start the particles to flow in the form of agglomerates that make the definition of an angle of repose much more difficult.

Using a shear test for calibration was not successful because of the noisy response of wetted systems when the test is conducted with a comparatively high pre-consolidation stress. This noisy behavior was consistent across both simulation and experiment. This agrees with literature, in which the resulting parameters of a shear test, namely the cohesion and the friction angle, did not vary consistently with the water loading, which is required to use the test for calibration.

In the developed workflow for the calibration of liquid bridge systems, the dry system is first calibrated. Subsequently, the minimum liquid layer height parameter and the surface tension are calibrated simultaneously based on the static angle of repose experiment. In the final step, the dynamic angle of repose test is used to calibrate the minimum separation distance parameter and the liquid viscosity. This segregated approach has the key advantage of calibrating only two parameters at a time, reducing the number of simulations required for the evaluation of a parameter set in each step. The applicability of this approach was successfully demonstrated for a mono-disperse glass particle system wetted with water.

In fluidized bed layering spray granulation, the final properties of the product depend on the microprocesses that occur either on the surface of the particles or within the droplets that end up on the particles' surface. Due to the complexity of the physics involved, an indirect particle property-prediction approach using tracked per-particle quantities was developed. The approach was able to derive high-fidelity mappings between particle properties and the tracked quantities for lab-scale granulation investigations using salt solutions and suspensions as spray liquid.

Applying this method to derive mappings to the experimental data by Schmidt et al. (2017b) who injected a suspension of limestone onto γ-alumina core particles, yielded a robust correlation between the target property of the work, the porosity of the coating shell, and a set of three parameters. The porosity of the granules decreased with higher mean liquid layer evaporation times and a higher mean droplet impact velocity while the porosity increased with higher solids concentrations in the droplets upon impact.

For spray granulation of salt solutions, own granulation experiments, published by Orth et al. (2021), were conducted by injecting a sodium benzoate solution onto Geldart group B micro-crystalline cellulose particles (Cellets). The surface roughness of the granulation products increased with lower fluidization air temperatures and higher liquid spray rates while it decreased with higher spray atomization pressure. The fluidization air flow rate and the atomization air temperature were not found to have a significant influence on surface roughness. As for the tracked quantities, a set of five statistical moments of the liquid layer evaporation time, the solids concentration in droplets upon impact and droplet impact velocity were found to suffice to map the local process conditions that droplets experience to the resulting surface roughness.

The predictiveness of the method in scale-up was demonstrated by scaling the aforementioned salt solution granulation scenario from the laboratory scale by a factor of eight to the pilot-scale. The bed height in the system, temperature, net spray rate per distributor area and gas velocity were kept constant between laboratory and pilot-scale experiments. The pilot-scale plant was equipped with four nozzles. By injecting the same liquid spray rate through either four or two of the nozzles, the interplay of wetting, mixing and evaporation was investigated. The resulting target product property, the surface roughness, was consistently higher in the pilot-scale experiments than in the laboratory-scale ones. This finding agrees with the observation in the laboratory-scale, that wetter drying conditions and lower temperatures produce particles with rougher surfaces. The application of the tracked quantity method predicted these deviations quantitatively when applying the derived mapping to the pilot-scale plant.

Based on the foundations laid in this work, further work on the liquid bridge state model is recommended. A comparison of the particle-scale simulations with resolved liquid distribution on the particles' surface like in the work of Schmelzle et al. (2018) would give a basis for further refinement. For extending the applicability to large-scale systems, a coarse-grained version of this model should be developed. To this end, a scaling study of the shear behavior should be sufficient. With this in place, describing the stability of fluidized beds with liquid injection quantitatively should be within reach of the method.

For the tracked quantity-particle property prediction approach, further applicability studies to polymer coatings and melts are recommended. By comparing the sensitivities of the tracked quantities, insights into the contributions of individual structure-forming microprocesses can be gained. Studies of geometric variants like top-spray and spouted bed granulators are advised to find general rules about the design of granulator with respect to desired product properties for a set of given scenarios. For example, the distance between spray nozzle and bed can be optimized to influence droplet drying to the degree which gives desired product properties for the layering granulation of salts, or the effect of additional aeration jets in pilot-scale plants to improve mixing can be investigated.

Finally, the tracked quantities recorded for individual particles could be used as an input for intra-particle scale resolution, for example with pore network modelling (Metzger et al., 2007), to augment or even replace the calibration step by means of first-principles modeling.

Bibliography

Ai, J., Chen, J.-F., Rotter, J. M., & Ooi, J. Y. (2011). Assessment of rolling resistance models in discrete element simulations. *Powder Technology*, *206*(3), 269–282.

Aigner, A., Schneiderbauer, S., & Kloss, C. (2013). Determining the Coefficient of Friction by Shear Tester Simulation, 9.

Antonyuk, S., Heinrich, S., Deen, N., & Kuipers, H. (2009). Influence of liquid layers on energy absorption during particle impact. *Particuology*, *7*(4), 245–259.

Bai, C., & Gosman, A. D. (1996). Mathematical Modelling of Wall Films Formed by Impinging Sprays. *SAE International Journal of Engines*, *105*(3).

Beetstra, R., van der Hoef, M. A., & Kuipers, J. A. M. (2007). Drag force of intermediate Reynolds number flow past mono- and bidisperse arrays of spheres. *AIChE Journal*, *53*(2), 489–501.

Benyahia, S., & Galvin, J. E. (2010). Estimation of Numerical Errors Related to Some Basic Assumptions in Discrete Particle Methods. *Industrial and Engineering Chemistry Research*, *49*(21), 10588–10605.

Bierwisch, C., Kraft, T., Riedel, H., & Moseler, M. (2009). Three-dimensional discrete element models for the granular statics and dynamics of powders in cavity filling. *Journal of the Mechanics and Physics of Solids*, *57*(1), 10–31.

Boyce, C. (2018). Gas-solid fluidization with liquid bridging: A review from a modeling perspective. *Powder Technology*, *336*, 12–29.

Boyce, Ozel, A., Kolehmainen, J., & Sundaresan, S. (2017). Analysis of the effect of small amounts of liquid on gas-solid fluidization using CFD-DEM simulations. *AIChE Journal*, *63*(12), 5290–5302.

Buck, B., Lunewski, J., Tang, Y., Deen, N. G., Kuipers, J., & Heinrich, S. (2018). Numerical investigation of collision dynamics of wet particles via force balance. *Chemical Engineering Research and Design*, *132*, 1143–1159.

Buck, B., Tang, Y., Heinrich, S., Deen, N. G., & Kuipers, J. (2017). Collision dynamics of wet solids: Rebound and rotation. *Powder Technology*, *316*, 218–224.

Burcat, A. (1984). Thermochemical Data for Combustion Calculations. *Combustion Chemistry*. Springer.

Burcat, A., & Ruscic, B. (2005). *Third Millennium Ideal Gas and Condensed Phase Thermochemical Database for Combustion with Updates from Active Thermochemical Tables* (TAE 960). Argonne National Laboratory.

Canny, J. (1986). A computational approach to edge detection. *IEEE Transactions on Pattern Analysis and Machine Intelligence*, *8*(6), 679–698.

Chan, E. L., & Washino, K. (2018). Coarse grain model for DEM simulation of dense and dynamic particle flow with liquid bridge forces. *Chemical Engineering Research and Design*, *132*, 1060–1069.

Chua, K. W., Makkawi, Y. T., Hewakandamby, B. N., & Hounslow, M. J. (2011). Time scale analysis for fluidized bed melt granulation-II: Binder spreading rate. *Chemical Engineering Science*, *66*(3), 327–335.

Clarke, A., Blake, T. D., Carruthers, K., & Woodward, A. (2002). Spreading and Imbibition of Liquid Droplets on Porous Surfaces. *Langmuir*, *18*(8), 2980–2984.

Coetzee, C. (2017). Review: Calibration of the discrete element method. *Powder Technology*, *310*, 104–142.

Combarros, M. (2014). Segregation of particulate solids: Experiments and DEM simulations. *Particuology*, *12*, 25–32.

Cundall, P. A., & Strack, O. D. L. (1979). A discrete numerical model for granular assemblies. *Géotechnique*, *29*(1), 47–65.

Dadkhah, M., Peglow, M., & Tsotsas, E. (2012). Characterization of the internal morphology of agglomerates produced in a spray fluidized bed by X-ray tomography. *Powder Technology*, *228*, 349–358.

de Souza Lima, R., Ré, M.-I., & Arlabosse, P. (2020). Drying droplet as a template for solid formation: A review. *Powder Technology*, *359*, 161–171.

Diez, E., Meyer, K., Bück, A., Tsotsas, E., & Heinrich, S. (2018). Influence of process conditions on the product properties in a continuous fluidized bed spray granulation process. *Chemical Engineering Research and Design*, *139*, 104–115.

Dosta, M., Antonyuk, S., & Heinrich, S. (2013). Multiscale Simulation of Agglomerate Breakage in Fluidized Beds. *Industrial & Engineering Chemistry Research*, *52*(33), 11275–11281.

Dosta, M., & Skorych, V. (2020). MUSEN: An open-source framework for GPU-accelerated DEM simulations. *SoftwareX*, *12*, 100618.

Ergun, S., & Orning, A. A. (1949). Fluid Flow through Randomly Packed Columns and Fluidized Beds. *Industrial & Engineering Chemistry*, *41*(6), 1179–1184.

Fries, L., Antonyuk, S., Heinrich, S., Niederreiter, G., & Palzer, S. (2014). Product design based on discrete particle modeling of a fluidized bed granulator. *Particuology*, *12*, 13–24.

Geometrical Product Specifications (GPS) - Surface texture: Profile method - Terms, definitions and surface texture parameters (Norm EN ISO 4287:1998). (2010). Beuth. Berlin OCLC: 845638068.

Gidaspow, D. (1994). *Multiphase Flow and Fluidization: Continuum and Kinetic Theory Descriptions* (1st ed.). Academic Press.

Girardi, M., Radl, S., & Sundaresan, S. (2016). Simulating wet gas–solid fluidized beds using coarse-grid CFD-DEM. *Chemical Engineering Science, 144*, 224–238.

Goniva, C., Kloss, C., Deen, N. G., Kuipers, J. A., & Pirker, S. (2012). Influence of rolling friction on single spout fluidized bed simulation. *Particuology, 10*(5), 582–591.

Gunn, D. (1978). Transfer of heat or mass to particles in fixed and fluidised beds. *Int. J. Heat Mass Transf., 21*(4), 467–476.

Han, Z., Xu, Z., & Trigui, N. (2000). Spray/Wall interaction models for multidimensional engine simulation. *International Journal of Engine Research, 1*(1), 127–146.

Härtl, J., & Ooi, J. Y. (2008). Experiments and simulations of direct shear tests: Porosity, contact friction and bulk friction. *Granular Matter, 10*(4), 263–271.

Henry G. Weller. (2021, June 13). *OpenFOAM-6.*

Hill, R. J., Koch, D. L., & Ladd, A. J. C. (2001). Moderate-Reynolds-number flows in ordered and random arrays of spheres. *Journal of Fluid Mechanics, 448*, 243–278.

Hoffmann, T. (2016). *Experimentelle Untersuchungen des Einflusses von verschiedenen Prozessparametern auf das Partikelwachstum bei der Wirbelschicht-Sprühgranulation.* OVGU Magdeburg. Germany.

Hoffmann, T., Rieck, C., Bück, A., Peglow, M., & Tsotsas, E. (2015). Influence of Granule Porosity during Fluidized Bed Spray Granulation. *Procedia Engineering, 102*, 458–467.

Jiang, Z., Rieck, C., Bück, A., & Tsotsas, E. (2020). Modeling of inter- and intra-particle coating uniformity in a Wurster fluidized bed by a coupled CFD-DEM-Monte Carlo approach. *Chemical Engineering Science, 211*, 115289.

Kafui, D., & Thornton, C. (2008). Fully-3D DEM simulation of fluidised bed spray granulation using an exploratory surface energy-based spray zone concept. *Powder Technology, 184*(2), 177–188.

Kariuki, W. I., Freireich, B., Smith, R. M., Rhodes, M., & Hapgood, K. P. (2013). Distribution nucleation: Quantifying liquid distribution on the particle surface using the dimensionless particle coating number. *Chemical Engineering Science, 92*, 134–145.

Kato, Z., Tanaka, S., Uchida, N., & Uematsu, K. (2006). Observation of the granule packing structure using a confocal laser-scanning microscope. *Journal of the European Ceramic Society, 26*(4-5), 683–687.

Kieckhefen, P., Lichtenegger, T., Pietsch, S., Pirker, S., & Heinrich, S. (2018a). Simulation of spray coating in a spouted bed using recurrence CFD. *Particuology*.

Kieckhefen, P., Pietsch, S., Dosta, M., & Heinrich, S. (2020). Possibilities and Limits of Computational Fluid Dynamics–Discrete Element Method Simulations in Process Engineering: A Review of Recent Advancements and Future Trends. *Annual Review of Chemical and Biomolecular Engineering, 11*(1), 397–422.

Kieckhefen, P., Pietsch, S., Höfert, M., Schönherr, M., Heinrich, S., & Kleine Jäger, F. (2018b). Influence of gas inflow modelling on CFD-DEM simulations of three-dimensional prismatic spouted beds. *Powder Technology*, *329*, 167–180.

Kloss, C., Goniva, C., Hager, A., Amberger, S., & Pirker, S. (2012). Models, algorithms and validation for opensource DEM and CFD-DEM. *Prog. Comput. Fluid Dyn. Int. J.*, *12*(2/3), 140.

Kolakaluri, R. (2013). *Direct Numerical Simulations and Analytical Modeling of Granular Filtration* (PhD thesis). Iowa State University. Ames, Iowa, United States of America.

Lambert, P., Chau, A., Delchambre, A., & Régnier, S. (2008). Comparison between Two Capillary Forces Models. *Langmuir*, *24*(7), 3157–3163.

Li, J., & Kwauk, M. (1994). *Particle-fluid two-phase flow: The energy-minimization multi-scale method*. Metallurgical Industry Press.

Lian, G., Thornton, C., & Adams, M. J. (1993). A Theoretical Study of the Liquid Bridge Forces between Two Rigid Spherical Bodies. *Journal of Colloid and Interface Science*, *161*, 138–147.

Lommen, S., Schott, D., & Lodewijks, G. (2014). DEM speedup: Stiffness effects on behavior of bulk material. *Particuology*, *12*, 107–112.

Lu, L., Yoo, K., & Benyahia, S. (2016). Coarse-Grained-Particle Method for Simulation of Liquid–Solids Reacting Flows. *Industrial and Engineering Chemistry Research*, *55*(39), 10477–10491.

Merson, E., Danilov, V., Merson, D., & Vinogradov, A. (2017). Confocal laser scanning microscopy: The technique for quantitative fractographic analysis. *Engineering Fracture Mechanics*, *183*, 147–158.

Metzger, T., Kwapinska, M., Peglow, M., Saage, G., & Tsotsas, E. (2007). Modern Modelling Methods in Drying. *Transport in Porous Media*, *66*(1-2), 103–120.

Mikami, T., Kamiya, H., & Horio, M. (1998). Numerical simulation of cohesive powder behavior in a fluidized bed. *Chemical Engineering Science*, *53*(10), 1927–1940.

Nasato, D. S. (2016). *Die Filling of Cohesive Powders: Material Characterization, Numerical Simulation and Experimental Validation*. Johannes-Kepler-University Linz. Linz, Austria.

Nase, S. T., Vargas, W. L., Abatan, A. A., & McCarthy, J. (2001). Discrete characterization tools for cohesive granular material. *Powder Technology*, *116*(2-3), 214–223.

Nijssen, T. M., Kuipers, H. A., van der Stel, J., Adema, A. T., & Buist, K. A. (2020). Complete liquid-solid momentum coupling for unresolved CFD-DEM simulations. *International Journal of Multiphase Flow*, *132*, 103425.

Noyes, A. A., & Whitney, W. R. (1897). The rate of solution of solid substances in their own solutions. *Journal of the American Chemical Society*, *19*(12), 930–934.

Ogarev, V. A., Timonina, T. N., Arslanov, V. V., & Trapeznikov, A. A. (1974). Spreading of Polydimethylsiloxane Drops on Solid Horizontal Surfaces. *Journal of Adhesion*, *6*(4), 337–355.

Orth, M., Kieckhefen, P., Pietsch, S., & Heinrich, S. (2021). *Correlating granule surface structure morphology and process conditions in fluidized bed layering spray granulation.*

Ozel, A., Kolehmainen, J., Radl, S., & Sundaresan, S. (2016). Fluid and particle coarsening of drag force for discrete-parcel approach. *Chemical Engineering Science*, *155*, 258–267.

Pareto, V. (1894). Il massimo di utilità dato dalla libera concorrenza. *Giornale degli Economisti*, *2*(9), 48–66.

Pedregosa, F., Varoquaux, G., Gramfort, A., Michel, V., Thirion, B., Grisel, O., Blondel, M., Prettenhofer, P., Weiss, R., Dubourg, V., Vanderplas, J., Passos, A., & Cournapeau, D. (2011). Scikit-learn: Machine Learning in Python. *Journal of Machine Learning Research*, *12*, 2825–2830.

Pietsch, S., Kieckhefen, P., Heinrich, S., Müller, M., Schönherr, M., & Kleine Jäger, F. (2018a). CFD-DEM modelling of circulation frequencies and residence times in a prismatic spouted bed. *Chemical Engineering Research and Design*, *132*, 1105–1116.

Pietsch, S., Kieckhefen, P., Müller, M., Schönherr, M., Kleine Jäger, F., & Heinrich, S. (2018b). Novel production method of tracer particles for residence time measurements in gas-solid processes. *Powder Technology*, *338*, 1–6.

Pirker, S., Kahrimanovic, D., & Goniva, C. (2011). Improving the applicability of discrete phase simulations by smoothening their exchange fields. *Applied Mathematical Modelling*, *35*(5), 2479–2488.

Pygall, S. R., Whetstone, J., Timmins, P., & Melia, C. D. (2007). Pharmaceutical applications of confocal laser scanning microscopy: The physical characterisation of pharmaceutical systems. *Advanced Drug Delivery Reviews*, *59*(14), 1434–1452.

Rabinovich, Y. I., Esayanur, M. S., & Moudgil, B. M. (2005). Capillary Forces between Two Spheres with a Fixed Volume Liquid Bridge: Theory and Experiment. *Langmuir*, *21*(24), 10992–10997.

Radl, S., Gonzales, B. C., Goniva, C., & Pirker, S. (2014). State of the Art in Mapping Schemes for Dilute and Dense Euler-Lagrange Simulations. *Progress in Applied CFD*, 9.

Radl, S., Radeke, C., Khinast, J. G., & Sundaresan, S. (2011). Parcel-based Approach for the Simulation of Gas-particle Flows. *Proceedings of the 8th International Conference on CFD in Oil & Gas, Metallurgical and Process Industries*, 124–134.

Richefeu, V., El Youssoufi, M. S., & Radjaï, F. (2006). Shear strength properties of wet granular materials. *Physical Review E*, *73*(5), 051304.

Rieck, C., Hoffmann, T., Bück, A., Peglow, M., & Tsotsas, E. (2015). Influence of drying conditions on layer porosity in fluidized bed spray granulation. *Powder Technology*, *272*, 120–131.

Roy, S., Singh, A., Luding, S., & Weinhart, T. (2016). Micro–macro transition and simplified contact models for wet granular materials. *Computational Particle Mechanics*, *3*(4), 449–462.

Salman, A., Hounslow, M., & Seville, J. (2006). *Granulation*. Elsevier Science.

Schmelzle, S., Asylbekov, E., Radel, B., & Nirschl, H. (2018). Modelling of partially wet particles in DEM simulations of a solid mixing process. *Powder Technology*, *338*, 354–364.

Schmidt, M., Bück, A., & Tsotsas, E. (2017a). Experimental investigation of the influence of drying conditions on process stability of continuous spray fluidized bed layering granulation with external product separation. *Powder Technology*, *320*, 474–482.

Schmidt, M., Bück, A., & Tsotsas, E. (2017b). Shell porosity in spray fluidized bed coating with suspensions. *Advanced Powder Technology*, *28*(11), 2921–2928.

Schneiderbauer, S., & Pirker, S. (2014). A Coarse-Grained Two-Fluid Model for Gas-Solid Fluidized Beds. *Journal of Computational Multiphase Flows*, *6*(1), 19.

Schneiderbauer, S., Kinaci, M. E., & Hauzenberger, F. (2020). Computational Fluid Dynamics Simulation of Iron Ore Reduction in Industrial-Scale Fluidized Beds. *steel research international*, *91*(12), 2000232.

Schulze, D. (2008). *Powders and bulk solids: Behavior, characterization, storage and flow*. Springer OCLC: ocn166372537.

Schulze, D. (2015). *Ring Shear Tester RST-XS.s Operating Instructions v1.2*. Manual. Dr. Dietmar Schulze Schüttgutmesstechnik.

Shi, D., & McCarthy, J. (2008). Numerical simulation of liquid transfer between particles. *Powder Technology*, *184*(1), 64–75.

Simons, T. A., Weiler, R., Strege, S., Bensmann, S., Schilling, M., & Kwade, A. (2015). A Ring Shear Tester as Calibration Experiment for DEM Simulations in Agitated Mixers – A Sensitivity Study. *Procedia Engineering*, *102*, 741–748.

Soulié, F., Cherblanc, F., El Youssoufi, M., & Saix, C. (2006). Influence of liquid bridges on the mechanical behaviour of polydisperse granular materials. *International Journal for Numerical and Analytical Methods in Geomechanics*, *30*(3), 213–228.

Stull, D. R. (1947). Vapor Pressure of Pure Substances. Organic and Inorganic Compounds. *Industrial & Engineering Chemistry*, *39*(4), 517–540.

Sutherland, W. (1893). The viscosity of gases and molecular force. *The London, Edinburgh, and Dublin Philosophical Magazine and Journal of Science*, *36*(223), 507–531.

Syamlal, M., & Gidaspow, D. (1985). Hydrodynamics of fluidization: Prediction of wall to bed heat transfer coefficients. *AIChE Journal*, *31*(1), 127–135.

Tausendschön, J., Kolehmainen, J., Sundaresan, S., & Radl, S. (2020). Coarse graining Euler-Lagrange simulations of cohesive particle fluidization. *Powder Technology*, *364*, 167–182.

Tenneti, S., Garg, R., & Subramaniam, S. (2011). Drag law for monodisperse gas–solid systems using particle-resolved direct numerical simulation of flow past fixed assemblies of spheres. *International Journal of Multiphase Flow*, *37*(9), 1072–1092.

Terrazas-Velarde, K., Peglow, M., & Tsotsas, E. (2009). Stochastic simulation of agglomerate formation in fluidized bed spray drying: A micro-scale approach. *Chemical Engineering Science, 64*(11), 2631–2643.

Tsuji, Y., Tanaka, T., & Ishida, T. (1992). Lagrangian numerical simulation of plug flow of cohesionless particle> in a horizontal pipe, 12.

Virtanen, P., Gommers, R., Oliphant, T. E., Haberland, M., Reddy, T., Cournapeau, D., Burovski, E., Peterson, P., Weckesser, W., Bright, J., van der Walt, S. J., Brett, M., Wilson, J., Millman, K. J., Mayorov, N., Nelson, A. R. J., Jones, E., Kern, R., Larson, E., Carey, C. J., Polat, İ., Feng, Y., Moore, E. W., VanderPlas, J., Laxalde, D., Perktold, J., Cimrman, R., Henriksen, I., Quintero, E. A., Harris, C. R., Archibald, A. M., Ribeiro, A. H., Pedregosa, F., & van Mulbregt, P. (2020). SciPy 1.0: Fundamental algorithms for scientific computing in Python. *Nature Methods, 17*(3), 261–272.

Wales, D. J., & Doye, J. P. K. (1997). Global Optimization by Basin-Hopping and the Lowest Energy Structures of Lennard-Jones Clusters Containing up to 110 Atoms. *The Journal of Physical Chemistry A, 101*(28), 5111–5116.

Wegner, S., Stannarius, R., Boese, A., Rose, G., Szabó, B., Somfai, E., & Börzsönyi, T. (2014). Effects of grain shape on packing and dilatancy of sheared granular materials. *Soft Matter, 10*(28), 5157.

Wilcox, D. C. (2006). *Turbulence modeling for CFD* (3rd ed). DCW Industries.

Willett, C. D., Adams, M. J., Johnson, S. A., & Seville, J. P. K. (2000). Capillary Bridges between Two Spherical Bodies. *Langmuir, 16*(24), 9396–9405.

Zang, D., Tarafdar, S., Tarasevich, Y. Y., Dutta Choudhury, M., & Dutta, T. (2019). Evaporation of a Droplet: From physics to applications. *Physics Reports, 804*, 1–56.

Zhou, Z. Y., Kuang, S. B., Chu, K. W., & Yu, A. B. (2010). Discrete particle simulation of particle–fluid flow: Model formulations and their applicability. *Journal of Fluid Mechanics, 661*, 482–510.

Zhu, H., Zhou, Z., Yang, R., & Yu, A. (2008). Discrete particle simulation of particulate systems: A review of major applications and findings. *Chemical Engineering Science, 63*(23), 5728–5770.

A CFD-DEM Modelling: Verification

The scope of this verification is restricted to spray deposition, vapor transport, heat transport and surface film evaporation. Verification of spray evaporation and heat transfer to droplets will not be performed as they are deemed mature and widely used in the context of combustible sprays.

A.1. Test Case Setup

Goal of this test case is to show conservation of mass and energy during a simple process.

The test case consists of a hexahedron spanning $0 \leq x \leq 0.04$, $0 \leq y \leq 0.02$, $0 \leq z \leq 0.02$, with a CFD mesh divided 17 times in the x-direction and 9 times in the y- and z-directions. All boundaries were declared to be no slip with respect to velocity and zero gradient to all others. This enforces the transport equations to conserve mass and energy.

A particle was inserted at $(0.01, 0.01, 0.01)$, at an initial temperature of $T_{\mathrm{p},0} = 300\,\mathrm{K}$ at a density of $\rho_\mathrm{p} = 1000\,\mathrm{kg\,m^{-3}}$. Both gas and particle are set to be at rest. The initial composition of the gas is $x_{\mathrm{N_2}} = 1$ at a pressure of 101 325 Pa. Viscosity is calculated using a simple mixture approach from the Sutherland equation, the heat capacity from *JANAF* fourth-order polynomials. The fluid state is calculated using an ideal gas state equation and the energy was calculated using the internal energy. $M_{\mathrm{s},0} = 1 \cdot 10^{-6}\,\mathrm{kg}$ particle surface liquid was created at a temperature of $T_{\mathrm{s},0} = 300\,\mathrm{K}$.

For the DEM side, the material properties were chosen according to Table A.1, resembling a particle coated with water. To show conservation and accurate implementation of the balances, the evaporation rate was fixed at $\dot{M}_{\mathrm{sg}}^{\mathrm{evap}} = 1 \cdot 10^{-9}\,\mathrm{kg\,s^{-1}}$.

A.1.1. Mass Transport and Conservation

Mass conservation has to be shown between the Eulerian and Lagrangian frames of references. Figure A.1 shows the change in vapor mass added to the gas phase, calculated by the integral

$$\int_V \rho_\mathrm{g} \alpha_\mathrm{g} y_{\mathrm{H_2O}} dV, \tag{A.1}$$

and mass of surface liquid lost by the particle to match perfectly over time. This is expected since the choice in implicit source term for the Eulerian side, elaborated in section 5.5.1,

Tab. A.1.: Material Properties used on the DEM side of the single particle evaporation test case.

	Symbol	Value
Particle		
Density	ρ_p	$1000\,\mathrm{kg\,m^{-3}}$
Heat Capacity	C_p	$1000\,\mathrm{Jkg^{-1}K^{-1}}$
Liquid		
Density	ρ_s	$1000\,\mathrm{kg\,m^{-3}}$
Heat Capacity	$C_{v,s}$	$4186\,\mathrm{J\,kg^{-1}\,K^{-1}}$
Heat Conductivity	k_s	$0.6\,\mathrm{W\,m^{-1}\,K^{-1}}$
Heat of Evaporation	Δh_s^{LV}	$2.5 \cdot 10^6\,\mathrm{J\,kg^{-1}}$

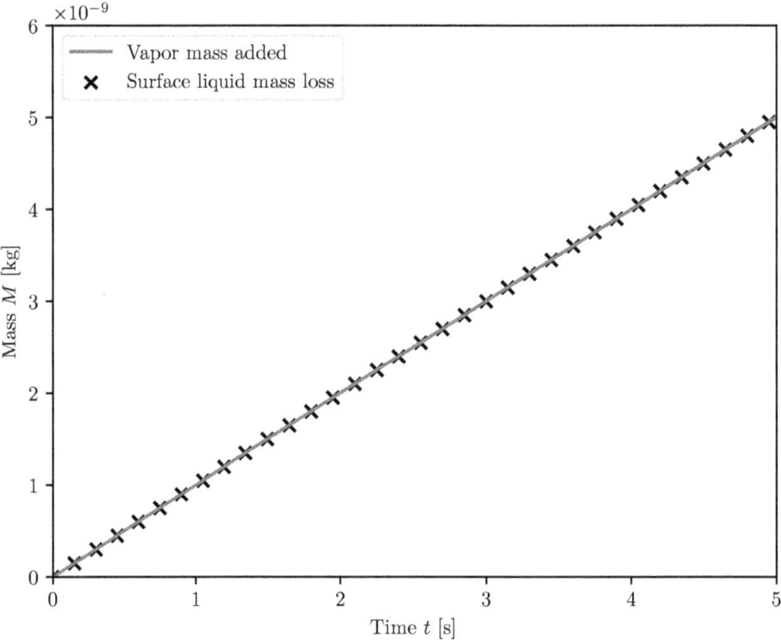

Fig. A.1.: Change in vapor mass and surface liquid mass over time at a fixed evaporation rate of $1 \cdot 10^{-9}\,\mathrm{kg\,s^{-1}}$.

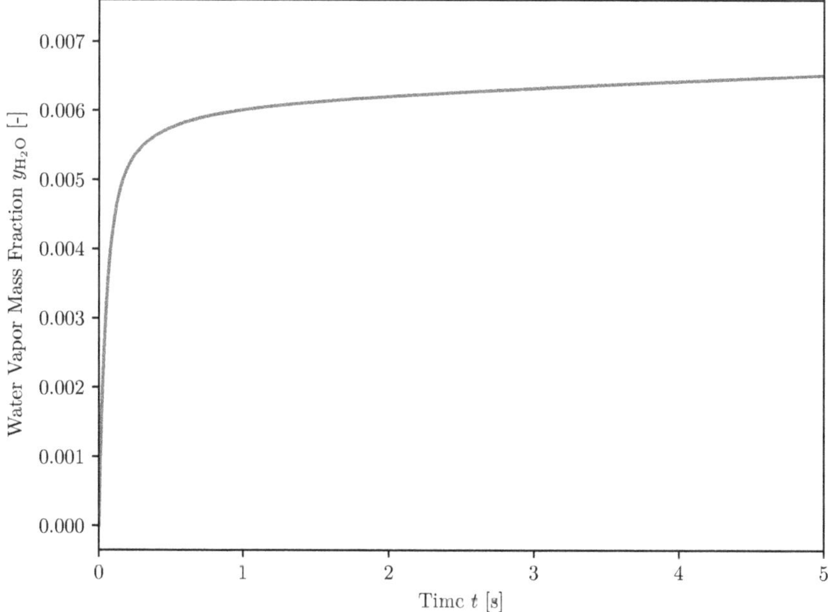

Fig. A.2.: Change in vapor mass fraction in the CFD grid cell containing the particle over time.

in combination with an explicit Lagrangian source term implies conservation. The change in vapor mass fraction in the cell containing the particle, and thus the one affected by the vapor source, is shown in Figure A.2. A steep increase in vapor concentration in this cell can be observed for the first 0.3 s, followed by a much slower steady, almost linear increase in mass fraction. This behavior shows that the interplay of a constant source term and the diffusion taking place is correctly implemented.

A.1.2. Energy Conservation

After showing mass conservation, energy conservation is the second concern to be addressed. For this purpose, the energies in the surface liquid $E_s = M_s C_s (T_s - T_{ref})$, the gas phase

$$E_g = \int_V \rho_g \alpha_g (C_{v,g}(T_g - T_{ref}) + y_{H_2O} \Delta h_s^{LV}) dV \quad \text{(A.2)}$$

and the particle phase $E_p = M_p C_p (T_p - T_{ref})$ must add to a constant value. Their changes relative to their initial value are plotted for the trial case in Figure A.3. As expected, the internal energy of the gas phase increases linearly due to the constant evaporation

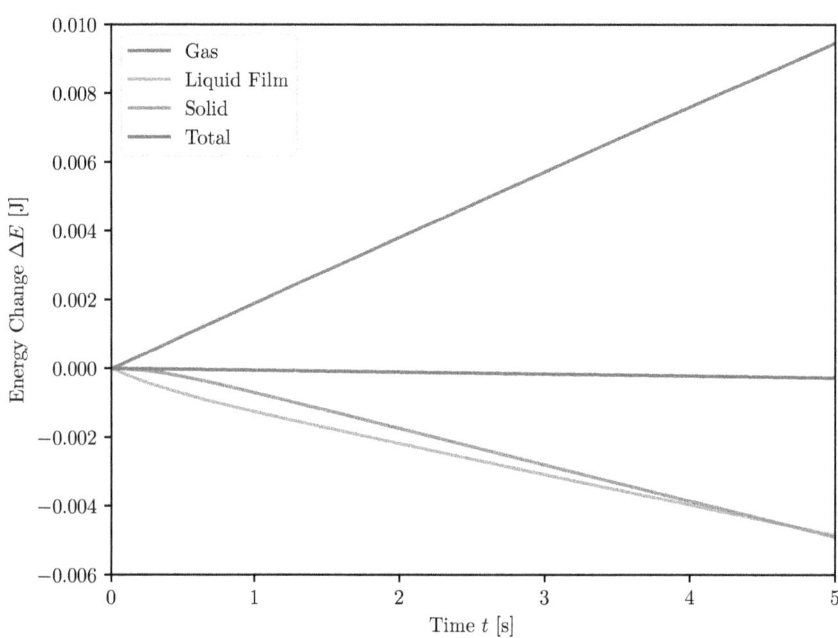

Fig. A.3.: Change in the internal energy of the gas phase, the surface liquid film, the particles and the entirety of the system. The evaporation rate was fixed at $1 \cdot 10^{-9}\,\mathrm{kg\,s^{-1}}$ for this scenario.

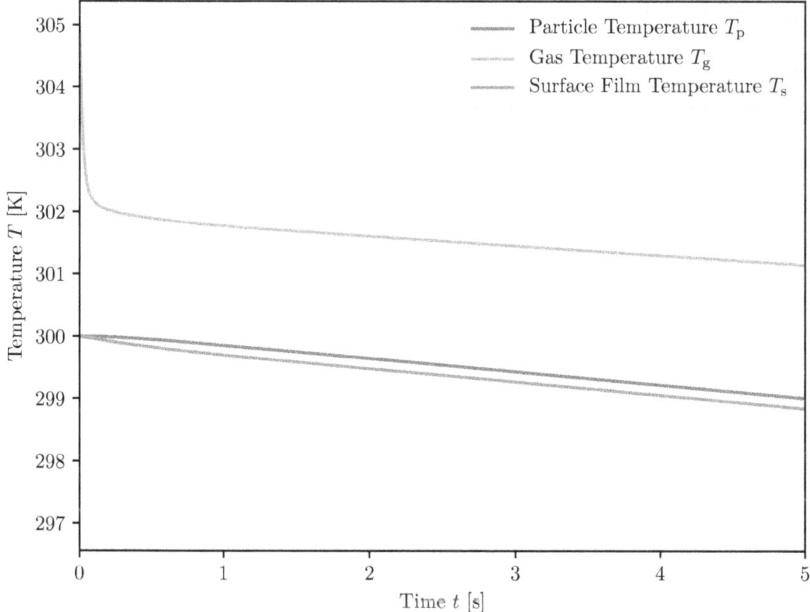

Fig. A.4.: Change in temperature in the single particle fixed-rate evaporation case.

rate. The behavior of the surface liquid film and solid is more varied, with the liquid film cooling down faster than the particle itself in the first 0.5 s due to the very high surface coverage and the higher heat capacity of the surface liquid. This can also be seen in the temperature profiles (Figure A.4) where the surface film temperature drops immediately and the change in particle temperature trails behind at a differential of approximately 0.1 K. Reason for this is the high heat exchange area between particle and surface liquid relative to their masses.

Solution Granulation Case: Supplementary Data

B

Tab. B.1.: Process conditions and resulting product properties for the Orth et al. (2021) case.

	Process Conditions							
	Fluidization Air		Spray	Atom. Air		Drying		
ID	Flowrate	Temp.	Rate	Pressure	Temp.	Potential	Avg. S_a	Ra
	$m^3\,h^{-1}$	°C	$g\,min^{-1}$	bar	°C	–	µm	
1	105	85	10	1.8	70	0.86	2.96	2.08
2	130	50	10	3	120	0.80	2.70	1.7
3	80	50	20	0.5	20	0.27	7.32	6.58
4	105	85	15	1.8	70	0.80	2.74	1.91
5	105	85	15	1.8	70	0.80	3.30	2.38
6	105	85	15	1.8	70	0.80	3.04	1.99
7	80	50	10	3	20	0.64	2.43	2.06
8	105	85	15	1.8	120	0.80	2.67	2.28
9	130	120	10	3	20	0.93	3.18	1.33
10	130	85	15	1.8	70	0.83	1.76	1.51
12	80	50	10	0.5	120	0.66	4.58	4.51
13	130	120	10	0.5	120	0.93	3.77	1.81
14	130	50	20	3	20	0.55	3.37	3.27
15	130	120	20	0.5	20	0.85	3.18	1.66
16	105	85	15	1.8	20	0.79	2.96	2.17
17	130	50	20	0.5	120	0.57	6.94	6.64
18	80	120	20	3	20	0.76	2.40	1.5
19	105	85	20	1.8	70	0.73	3.30	2.3
20	80	120	20	0.5	120	0.76	4.89	2.17
21	130	120	20	3	120	0.86	2.56	1.18
22	80	120	10	0.5	20	0.88	3.83	2.07
23	105	85	15	3	70	0.80	3.28	1.8
24	130	50	10	0.5	20	0.78	4.04	3.08
25	105	120	15	1.8	70	0.86	2.65	1.87
26	80	120	10	3	120	0.89	2.93	1.38
28	105	50	15	1.8	70	0.61	4.15	3.9
30	80	85	15	1.8	70	0.73	2.98	3.02
31	80	50	20	3	120	0.38	3.26	2.48
32	105	85	15	0.5	70	0.79	3.20	2.38

Tab. B.2.: Process conditions and resulting tracked quantities for the Orth et al. (2021) case.

	Process Conditions						Tracked Quantities								
	Fluidization Air		Spray	Atom. Air		Drying	Evaporation Time			Solids Concentration			Impact Velocity		
ID	Flowrate	Temp.	Rate	Pressure	Temp.	Potential	μ_0	μ_1	μ_2	μ_0	μ_1	μ_2	μ_0	μ_1	μ_2
	$m^3 h^{-1}$	°C	$g\,min^{-1}$	bar	°C	–	s	s	-	$kg\,kg^{-1}$	$kg\,kg^{-1}$	-	$m\,s^{-1}$	$m\,s^{-1}$	-
1	105	85	10	1.8	70	0.86	3.15	4.50	2.12	0.49	0.13	1.01	1.14	0.81	1.69
2	130	50	10	3	120	0.80	4.94	5.01	1.61	0.50	0.11	1.11	1.17	0.82	1.76
3	80	50	20	0.5	20	0.27	2.49	2.84	3.06	0.31	0.01	2.93	1.00	0.71	2.53
4-6	105	85	15	1.8	70	0.80	3.15	4.27	2.05	0.39	0.07	2.16	1.59	1.04	1.84
7	80	50	10	3	20	0.64	4.88	4.50	1.42	0.41	0.09	1.95	0.95	0.8	2.18
8	105	85	15	1.8	120	0.80	3.30	4.74	2.19	0.40	0.08	2.08	1.61	1.04	1.80
9	130	120	10	3	20	0.93	3.47	5.38	2.28	0.55	0.15	0.40	1.79	1.03	1.33
10	130	85	15	1.8	70	0.83	2.99	4.26	2.18	0.38	0.06	2.37	1.67	1.02	1.62
12	80	50	10	0.5	120	0.66	3.74	3.91	1.61	0.40	0.07	2.24	0.87	0.72	2.29
13	130	120	10	0.5	120	0.93	2.32	4.28	2.68	0.57	0.14	0.44	1.43	0.85	1.31
14	130	50	20	3	20	0.55	3.77	3.39	1.38	0.34	0.02	2.11	1.54	1.06	1.79
15	130	120	20	0.5	20	0.85	1.62	2.96	2.79	0.44	0.11	1.35	1.47	0.87	1.56
16	105	85	15	1.8	20	0.79	2.76	3.78	2.11	0.37	0.06	2.47	1.44	0.91	1.80
17	130	50	20	0.5	120	0.57	3.04	3.37	1.77	0.33	0.01	1.29	1.36	0.82	1.87
18	80	120	20	3	20	0.76	2.59	3.83	2.21	0.48	0.15	0.90	1.59	1.16	1.77
19	105	85	20	1.8	70	0.73	2.97	4.03	2.09	0.38	0.06	2.32	1.45	0.94	1.89
20	80	120	20	0.5	120	0.76	1.59	2.72	2.57	0.47	0.12	1.12	1.25	0.88	1.98
21	130	120	20	3	120	0.86	3.30	5.06	2.28	0.51	0.14	0.80	1.91	1.14	1.61
22	80	120	10	0.5	20	0.88	2.27	4.03	2.63	0.54	0.15	0.54	1.08	0.81	1.60
23	105	85	15	3	70	0.80	3.52	4.36	1.84	0.44	0.11	1.53	1.56	1.05	1.79
24	130	50	10	0.5	20	0.78	3.17	3.31	1.64	0.38	0.06	2.56	1.04	0.72	1.61
25	105	120	15	1.8	70	0.86	2.47	4.20	2.48	0.43	0.11	1.59	1.66	1.02	1.65
26	80	120	10	3	120	0.89	3.99	6.71	2.45	0.51	0.13	0.92	1.91	1.1	1.51
28	105	50	15	1.8	70	0.61	4.42	4.61	1.66	0.33	0.02	1.69	1.45	0.9	1.85
30	80	85	15	1.8	70	0.73	3.49	4.80	2.05	0.39	0.07	2.12	1.45	0.97	1.82
31	80	50	20	3	120	0.38	5.40	4.75	1.26	0.35	0.02	1.80	1.34	0.97	2.01
32	105	85	15	0.5	70	0.79	2.14	3.18	2.31	0.39	0.07	2.05	1.34	0.85	1.81

Tab. B.3.: Principal components and eigenvalues of the surface roughness and tracked quantities.

	Roughness	Evaporation Time			Solids Concentration			Impact Velocity		
Eigenvalue	S_a	$\mu_0(t_{evap})$	$\mu_1(t_{evap})$	$\mu_2(t_{evap})$	$\mu_0(x_{s,imp})$	$\mu_1(x_{s,imp})$	$\mu_2(x_{s,imp})$	$\mu_0(v_{imp})$	$\mu_1(v_{imp})$	$\mu_2(v_{imp})$
1,553	0.24	0.05	−0.3	−0.13	−0.42	−0.41	0.38	−0.31	−0.28	0.4
900	−0.25	0.54	0.36	−0.49	−0.2	−0.21	0.16	0.23	0.32	0.02
528	0.05	0.4	0.21	−0.29	0.3	0.2	−0.3	−0.52	−0.47	0.05
285	−0.86	−0.15	−0.24	−0.07	0.02	0.16	0.28	−0.25	−0.11	−0.02
201	0.09	−0.13	−0.49	−0.55	−0.09	−0.15	−0.27	0.03	−0.02	−0.57
126	0.04	−0.01	−0.38	−0.26	0.1	0.32	−0.26	−0.06	0.5	0.59
59	−0.35	0.14	−0.1	0.29	−0.17	−0.47	−0.68	0.1	−0.11	0.19
14	0.05	0.63	−0.49	0.36	0.33	−0.08	0.21	−0.04	0.17	−0.21
9	−0.02	0.08	−0.2	−0.17	0.22	0.14	0.1	0.7	−0.54	0.26
5	0.03	0.28	−0.1	0.18	−0.7	0.59	−0.11	0.06	−0.11	−0.12

C Calculation of Thermophysical Properties in OpenFOAM

C.1. Calculation of the Density

Ideally, the density in the gas phase is calculated using the expression

$$\rho_g = \frac{p}{RMT_g}, \quad \text{(C.1)}$$

where ρ_g is the gas density, p is the pressure, $R = 8.314\,\mathrm{J\,mol^{-1}\,K^{-1}}$ is the ideal gas constant, $M_g = \sum_i y_i M_{g,i}$ is the mean molecular weight and T_g is the gas phase temperature.

In practice, the iterative nature of the OpenFOAM requires the consideration of advection. The details of the implementation can be found in the source code (Henry G. Weller, 2021).

C.2. Calculation of the Heat Capacity

According to Burcat (1984), the heat capacity is calculated using a five-coefficient polynomial

$$\frac{c_p(T)}{R} = a_1 + a_2T + a_3T^2 + a_4T^3 + a_5T^4, \quad \text{(C.2)}$$

where a_i are the aforementioned coefficients.

In this work, the coefficients from the authoritative work by Burcat and Ruscic (2005) are used, as reproduced in Tab. C.1.

Tab. C.1.: Coefficients for the calculation of thermodynamic state data used. Only the lower range of coefficients for the temperature range 200 K < T_g < 1000 K is shown.

Species	a_1	a_2	a_3	a_4	a_5	a_6
	-	K^{-1}	K^{-2}	K^{-3}	K^{-4}	K
N_2	3.53	−0.000 124	$-5.03 \cdot 10^{-7}$	$2.44 \cdot 10^{-9}$	$-1.41 \cdot 10^{-12}$	1050
H_2O	4.20	−0.002 04	$6.52 \cdot 10^{-6}$	$-5.49 \cdot 10^{-9}$	$1.77 \cdot 10^{-12}$	−30 300

C.3. Calculation of the Viscosity

According to Sutherland (1893), the temperature-dependence of the gas phase viscosity η_g can be modelled using the expression

$$\eta_g = \frac{A_s T_g^{1.5}}{T + T_s}, \tag{C.3}$$

where A_s and T_s are the Sutherland coefficients.

Curriculum Vitae

Personal information

Family name	Kieckhefen
First name	Paul
Citizenship	German
Date of birth	14.12.1992
Place/country of birth	Hamburg, Germany

Education

1997-2001	Katholische Schule Neugraben
2001-2011	Niels-Stensen-Gymnasium Hamburg
2011	Abitur

Academic history

10.2011-09.2013	Study of Molecular Life Sciences, University of Hamburg
10.2013-03.2016	Bachelor of Science in Bioprocess Engineering, Hamburg University of Technology
04.2016-01.2018	Master of Science in Process Engineering, Hamburg University of Technology
02.2018-09.2021	Research Assistant, Institute for Solids Process Engineering and Particle Technology, Hamburg University of Technology

Internships

04.2016-09.2016	Internship at BASF SE, Ludwigshafen am Rhein
09.2017-12.2017	Internship at BASF SE, Ludwigshafen am Rhein

Professional career

since 10.2021	Senior Specialist Digitalization, Computational Fluid and Particle Dynamics, BASF SE, Ludwigshafen am Rhein

www.ingramcontent.com/pod-product-compliance
Ingram Content Group UK Ltd.
Pitfield, Milton Keynes, MK11 3LW, UK
UKHW061826190726
13853UKWH00009B/2459

9 783736 975569